Rhizome

BY GWEN DIEHN

ACKNOWLEDGMENTS

Without the encouragement of on-line followers and friends this book
would not have come about, and so a big thanks to all of you who
believed these images and texts were worth pulling together and who
urged me to make a book. Thanks especially to Laura, one of my first
encouragers, who worked her magic and pulled my scattered images
and text files into a cohesive whole. Special thanks also to Maria
for gentle and patient editing of the introduction, and to Jorge for
suggestions regarding digital imaging and for insisting that the images
be the best they possibly could be.

Gwen Diehn's prints, drawings and artists' books have been exhibited
internationally and are in many collections, including the National
Museum of Women in the Arts, Washington, DC. She has taught art at
the college level and in short courses and workshops for many years.
She is the author of several Sterling/Lark books, including **Simple
Printmaking** (2000), **Real Life Journals (Journal Your Way)** paperback
title) (2010), and **The Complete Decorated Journals** (2013).

ISBN: 978-0-578-28080-6

Design and digital production by Laura Ladendorf

Green Turtle Books

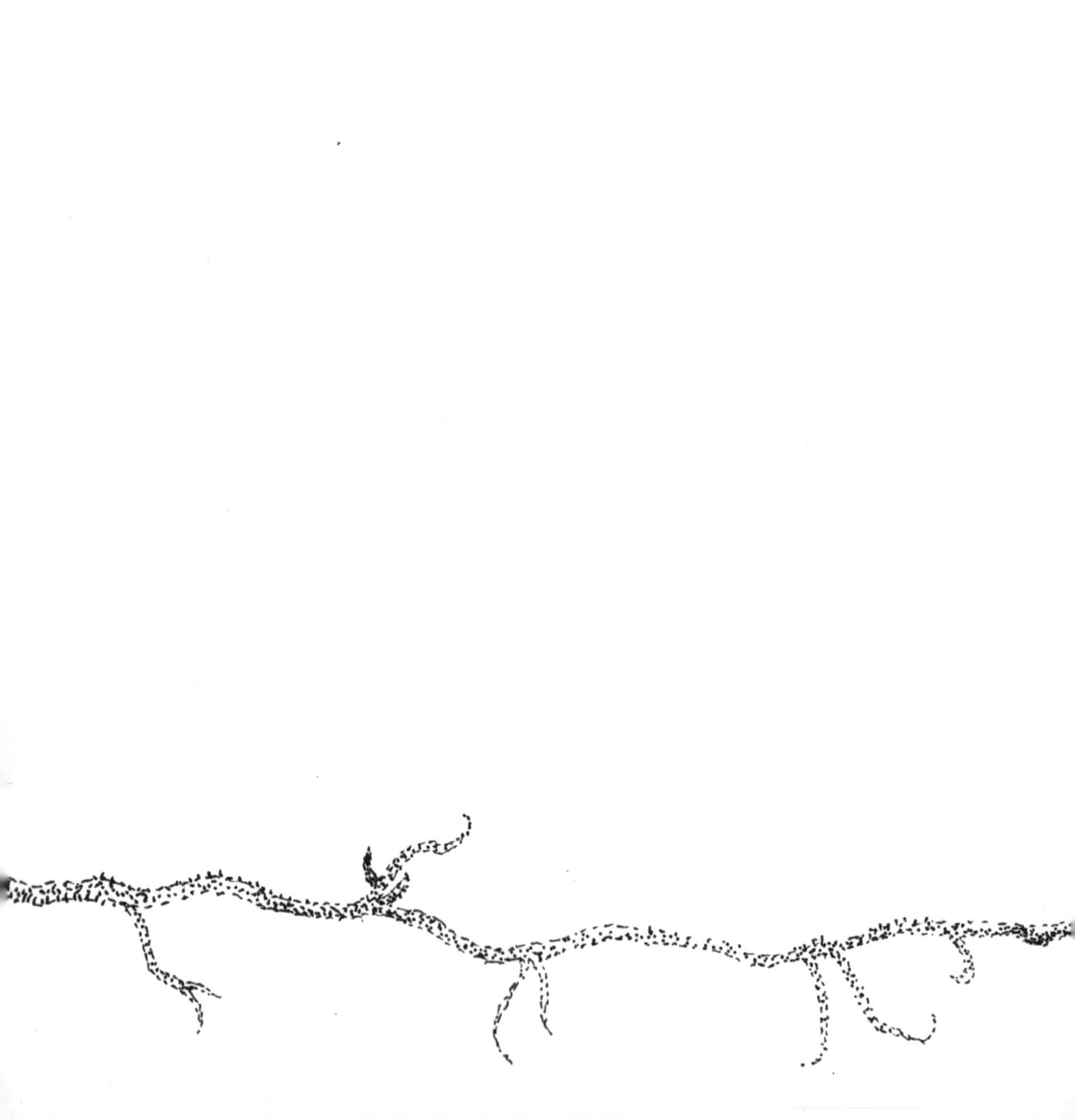

Solitary tree
Will not grow strong by itself –
Children feed the mom

Richard Greene

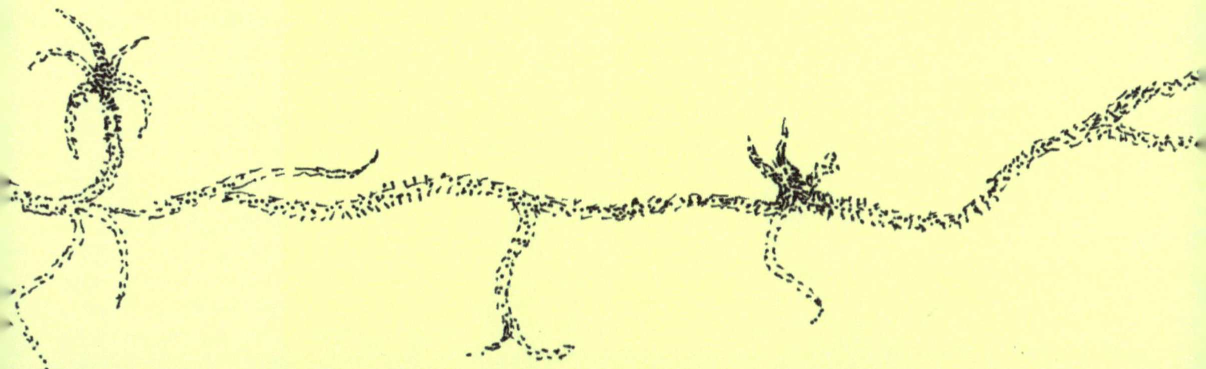

INTRODUCTION

This little book is a collection of posts that I made on social media over the past year. I had already had an established practice of almost daily drawing, and during the time of Covid lockdown, I spent a lot of time sketching outside on nearby trails that had been closed for most of the pre-vaccine year. The plants and bryophytes, insects and small mammals had flourished during the year of seclusion.

The book begins somewhere in a middle, and it meanders off in another middle, leaving some questions answered and others still floating:

> Where does the forsythia end and the bee begin? What is the thread that ties the bee to the sun to the forsythia to the honey to the human and the bear? Why are the first blooms before the end of winter reddish and why does the still-cold air not hurt these very early flowers? And then, why are so many of the next blooms yellow? Why are the enclosures of seeds so complex and various, and how do squirrels relate to these strange constructions? How do lichens come back to life after they have apparently dried up? Why are there blooms at the end of winter when there are almost no pollinators around?

At the back of the book are a few blank pages for you to make notes and sketches and some connections of your own. As such, this book about rhizomes both botanical and philosophical is itself a rhizome[1] more than a logically organized structure. You can open it at random, and meander through as you please. Occasionally a couple of posts will reference each other, but you won't get confused by reading the pages out of order, enjoying any connections that you can make. As with any rhizome, there are places from which new paths and shoots and roots can grow: you will find links to more information in most of the entries.

I began these drawings as an aid to looking and studying. Curiosity soon drove me to research and write about my observations as well as about the information I discovered. May the book open up some new ways of looking for you, too.

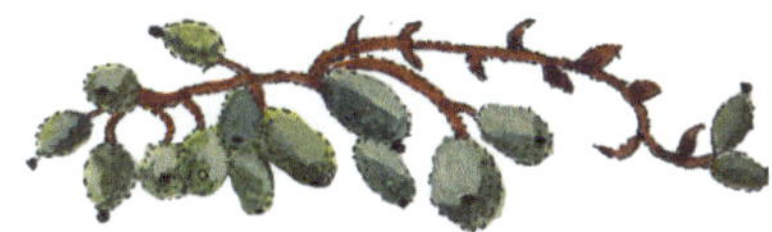

¹rhizome (in botany) – an underground, horizontal stem that in some ways functions like a root in that it anchors the plant to the ground, it takes up water from the soil, it stores nutrients and food made by photosynthesis; but a rhizome can also send up shoots that can develop into new plants, and it can also send down roots. It has no beginning or end point, and it ceaselessly seeks to establish new connections.

¹rhizome (in post-modern philosophy) – a philosophical concept used to describe systems with no clear beginning or end such as the internet. (Deleuze Giles. and Felix Guattari, A Thousand Plateaus: Capitalism and Schizophrenia, University of Minnesota Press, 1987, pages 11-19.)

This entire archive is available online using: 10000drawings on Instagram or Gwen Diehn on Facebook. Posts spanned May 4, 2021 through March 17, 2022. Many of the links are accessible as live links via Facebook postings.

FLAME AZALEA

The surviving flame azalea on our bush - this blossom managed to survive 24 degrees F temperatures, high winds, hot sun, and beating rain. This morning I went out to check on it and found it blooming. There are a few others on the bush, but they're battered and brown for the most part. This is a native plant here in the Appalachians, a member of the *Ericaceae* family (of which blueberries are an edible member). All parts of the flame azalea are poisonous to humans, but valuable for bees who pollinate it and eat the nectar.

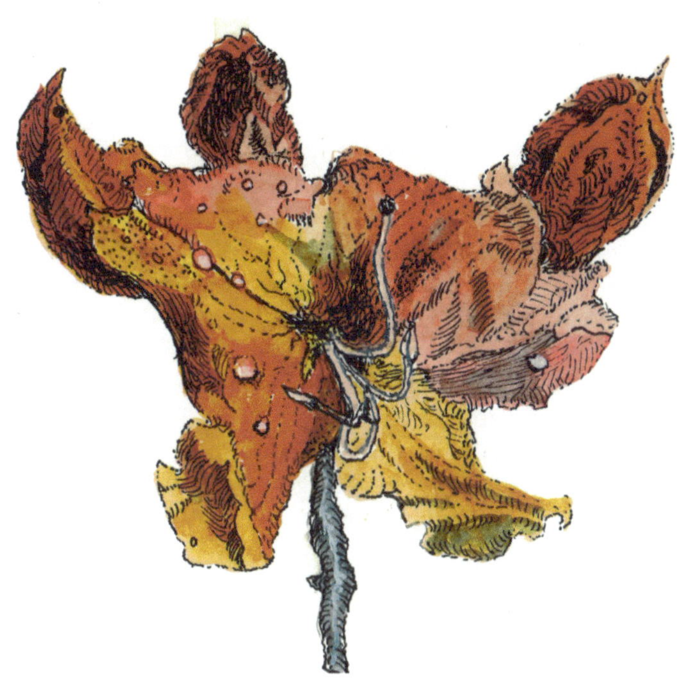

PAULOWNIA AND TULIP POPLAR

So high up they can't be seen until they get blown down, these paulownia (*Pauloniaceae* family) flowers and a tulip poplar tree bloom were on the ground in front of me today. Both are great pollinator plants providing nectar to bees. The paulownia, also called the princess tree, has been around this continent for at least 40,000 years, according to fossils found in Washington state. Around here people love to hate it, calling it invasive. The truth is that one kind (paulownia tomentosa) grows 20-40 feet a year and can take over a yard. But there are many species of paulownia; and two especially, P. fortuna and P. elongata, are prized by permaculturists especially for their leaves (green fertilizer) and nitrogen-fixing ability. The leaves provide over 20% protein and grow well in depleted soils. Paulownia trees have been in the Traditional Chinese Medicine pharmacopeia for centuries due to the anti-viral qualities of their roots.

Tulip trees provide pollen for honeybees and hummingbirds as well as bird and mammal feed. The straight trunks have been used for dugout canoes as well as railroad ties. I know from experience that their wood is excellent for woodblock carving.

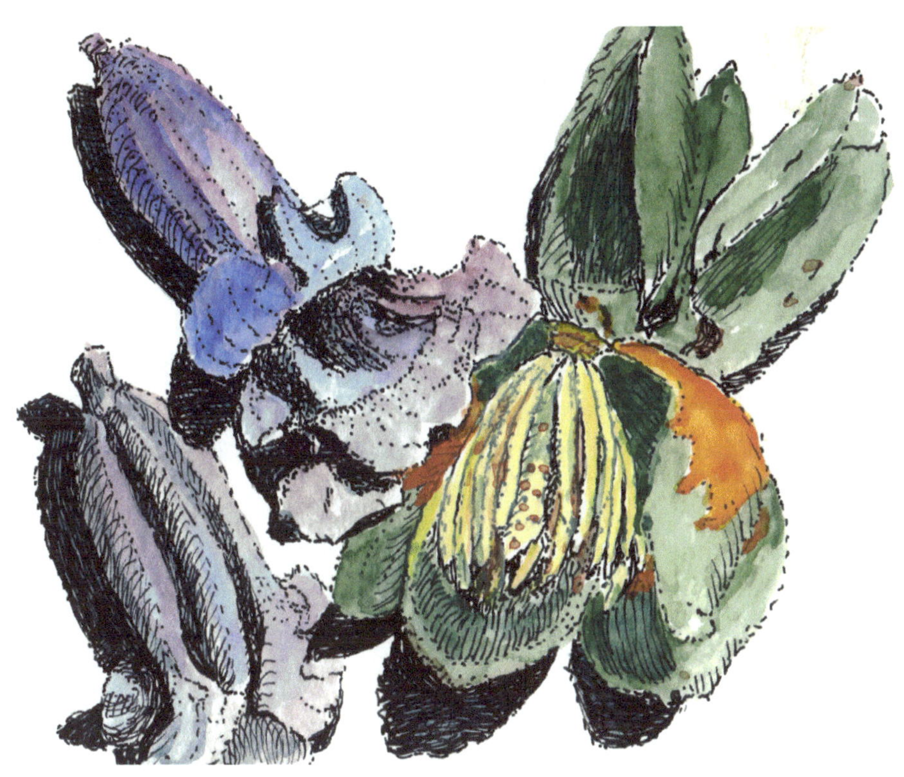

LEATHER LEAFED MAHONIA
(Oregon grape)

The leather leafed mahonia (*Berberidaceae* family) is an elaborate bush with unexpected gifts to give if you can get past its thorny, leathery leaves. For one thing, it blooms in winter, small yellow bells with a light, sweet scent (if you can poke your nose in between the thorns to sniff it); those flowers are a boon for early bees out looking in winter for what nectar they can find. The fruits are another gift. Birds love these greenish greyish bluish berries, which are quite nutritious.

Mahonia has been around for a long time. Fossils of it 20,000 years old have been found in Washington state in the US; Native People used it as medicine as well as a food source added to porridge and in the form of jellies and jams. The tart berries become softer and sweeter after weathering a freeze or two. On top of all that, the plant is beautiful in all seasons with its green-orange-violet and yellow colors. Our mahonia berries are still ripening, turning from green to purplish. For more details:

https://pfaf.org/user/Plant.aspx?LatinName=Mahonia+bealei

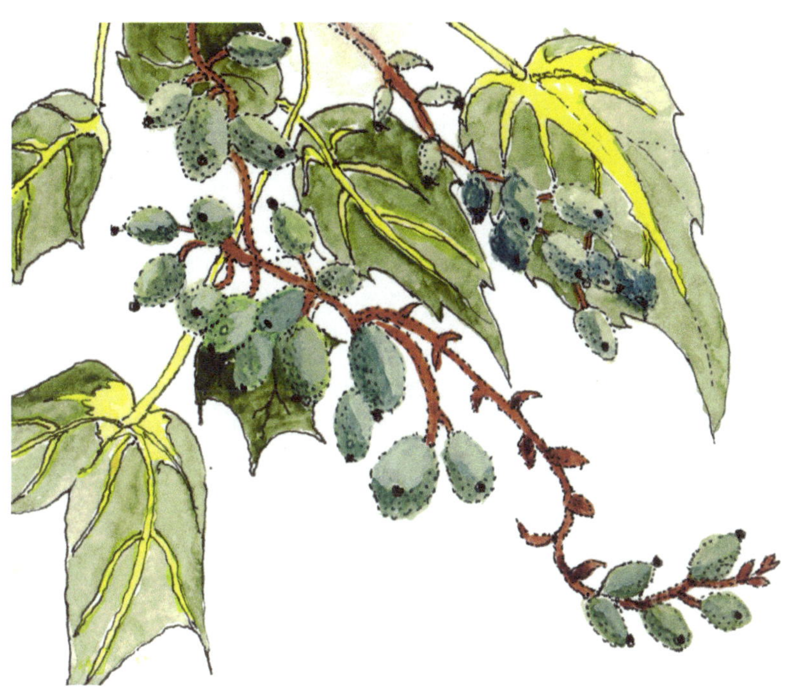

YELLOW CHESTNUT TREE -
buckeyes forming

Three buckeyes are forming in the yellow buckeye tree that I've been following. In order to grow they seem to need to shed former skins like snakes. Most of the pretty peanut-shaped flowers in the cluster fall off and land on the ground under the tree. You can see the little bumps where they were on the twigs. But those that have been pollinated and are able to burst out of their too-small skins can go on to develop into seeds. They remind me of small planets.

A couple of facts: the buckeye trees around here (aesculus glabrus) are in the *Sapindaceae* family of plants, and the alkaline saponins in every part of them are poisonous to humans as well as most animals and insects. Even so, squirrels and deer seem to be immune to the toxins as well as native honeybees and hummingbirds and some butterflies. Native people were able to leach the alkaline toxins from buckeyes by crushing and then roasting or boiling them and using them for flour as they used acorns. They also used the soapy foam that came out of the boiling buckeyes to stun or kill fish, which they could then easily scoop up.

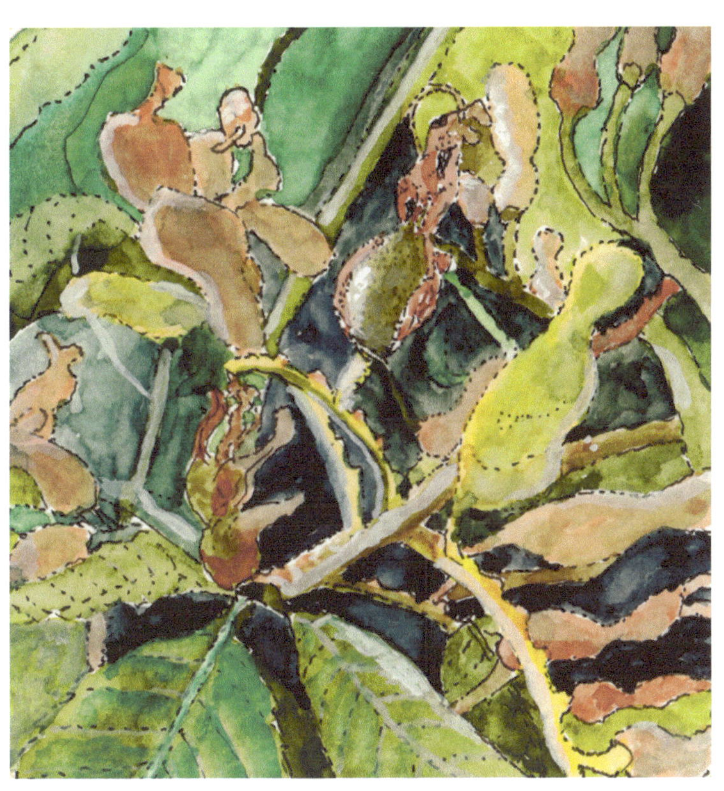

YELLOW CHESTNUT TREE -
buckeyes maturing

Peek into the heart of the yellow chestnut tree (*Sapondaceae* family), into the heart of its single blossom, where one of the multiple flowers has begun to transform into the seed, which we call a conker or buckeye! The flowers start out yellow, then pick up apricot-colored tinges. Many fall off the tree at this point, but the ones that have been pollinated begin to transform. The bottom part of the peanut-shaped flower swells and bursts out of its covering. The nascent buckeye at this point is greenish. You can still see shreds of its pinkish skin. It's in the upper right quadrant toward the center of the drawing. Stay tuned for further developments. This flower is right at my eye level on a tree on the edge of the river trail.

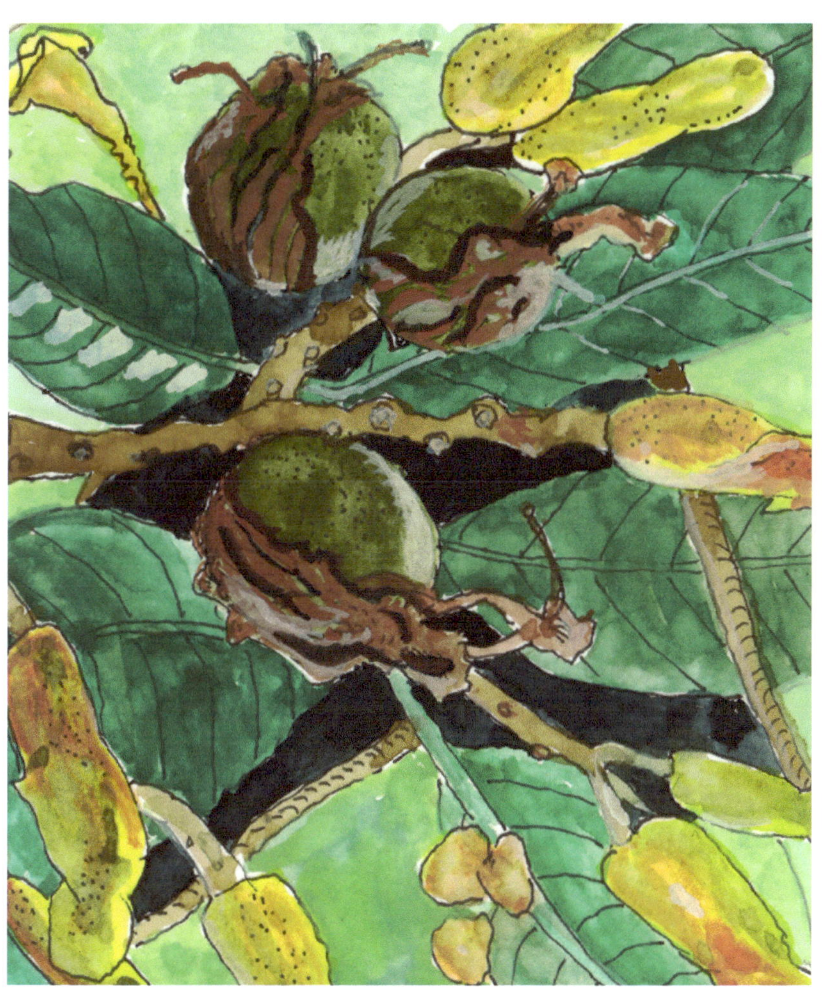

LYRE LEAF SAGE

Lyre Leaf Sage is all along the trails and in many of the yards on our road. It's in the mint family *(Lamiaceae)*, which you can easily tell by its square stem. The entire plant can be dried and made into a tea that not only tastes good but has been used medicinally for colds, coughs, and sore throats. Native people made a salve out of the roots that will help heal sores. On top of all that, salvia lyrate attracts hummingbirds, birds, and butterflies!

https://plants.ces.ncsu.edu/plants/salvia-lyrata/

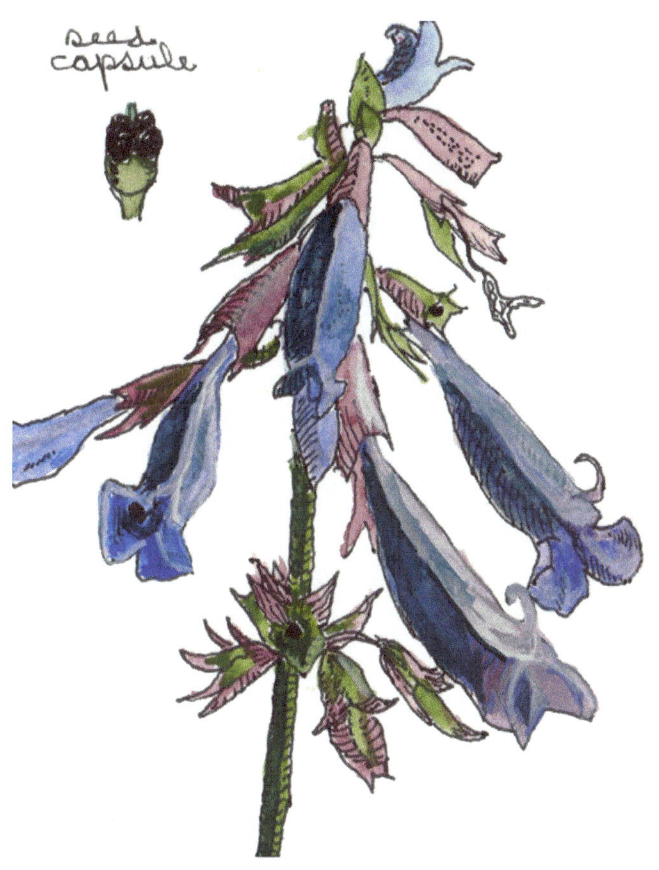
seed
capsule

SOLOMON'S SEAL

Here is Solomon's Seal, seen from an unusual view. Solomon's seal (*Asparagaceae* family) is very commonplace in our woods, but it's always a lovely biennial surprise to see the big frond with the marching pairs of dangling bell-shaped flowers. Pollinated by bumblebees and butter-flies, it also feeds birds when its black berries appear in later summer. The rhizome of Solomon's Seal has been used for thousands of years as medicine as well as food. Called huang jing in Chinese medicine, it is used as a yin tonic to benefit the liver, kidney and spleen. The Chippewa sprinkled a decoction of the roots on hot stones as a remedy for head-aches. (An interesting connection with traditional Chinese Medicine's use of it as a yin tonic, which helps keep energy or qi from rising, which often causes headaches.)

Solomon's Seal is also useful as an astringent to heal bruises, edema, hemorrhoids, ulcers, and boils and to treat lung disorders, inflammation, and to reduce swelling.

Early Colonists ate the rhizome boiled like potatoes, which they no doubt learned from Native people, who used the root like potatoes and also in soups. They also boiled young shoots and ate them like asparagus.

And if all the above isn't enough, Solomon's Seal is a main ingredient in King Solomon's Oil, an essential oil of hyssop, Solomon's seal, bay, frankincense, and rose in a sweet almond oil carrier; this formula is said to improve mental clarity and wisdom as well as to draw success and protection as well as removal of jinxes.

https://en.wikipedia.org/wiki/Polygonatum

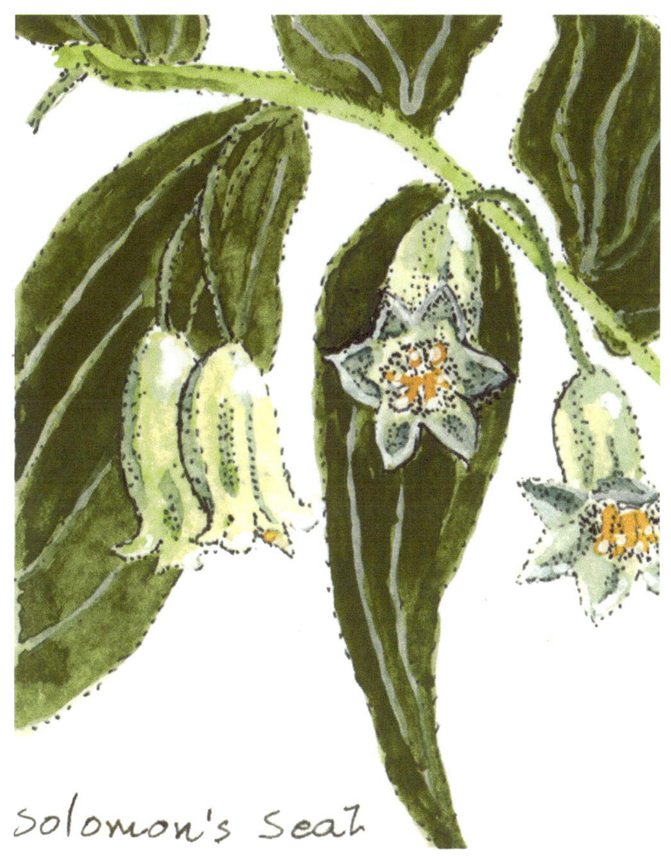

Solomon's Seal

BLACKBERRY BLOOMS

Two ancient medicinal/food plants are on my radar today. In the *Rosaceae* family is blackberry, which is suddenly everywhere in our yard, on the trails, and along our road. Blackberries were found in the stomach of the remains of the 2500-year-old Haraldskaer Woman in a bog in Denmark. And the 1696 London Pharmacopeia describes the use of blackberry cordial for curing stomach ulcers.

The drawing shows the formation of the berry from the compound flowers of the plant. Each berry is a cluster of seeds surrounded by their ripening uteruses, all teeming with antioxidants, anti-irritants, vitamins, and fiber.

On the right are three stems of horsetail, which I found growing in a marshy area near here. Horsetail is believed to be one of the earliest plants. It has been used for thousands of years because of the silica in its stem and rhizome and leaves. Not only is it a fun plant to play with (you can bend it at its stripey joints, and you can literally watch it grow), but it is used to treat achy joints and osteoporosis as well as weak, brittle nails and hair. It is usually taken as a tea, but can be mashed and directly applied to wounds, burns, and frostbite. It is used in Traditional Chinese Medicine as well as in Western medicine. Research supports most of its traditionally- experienced benefits.

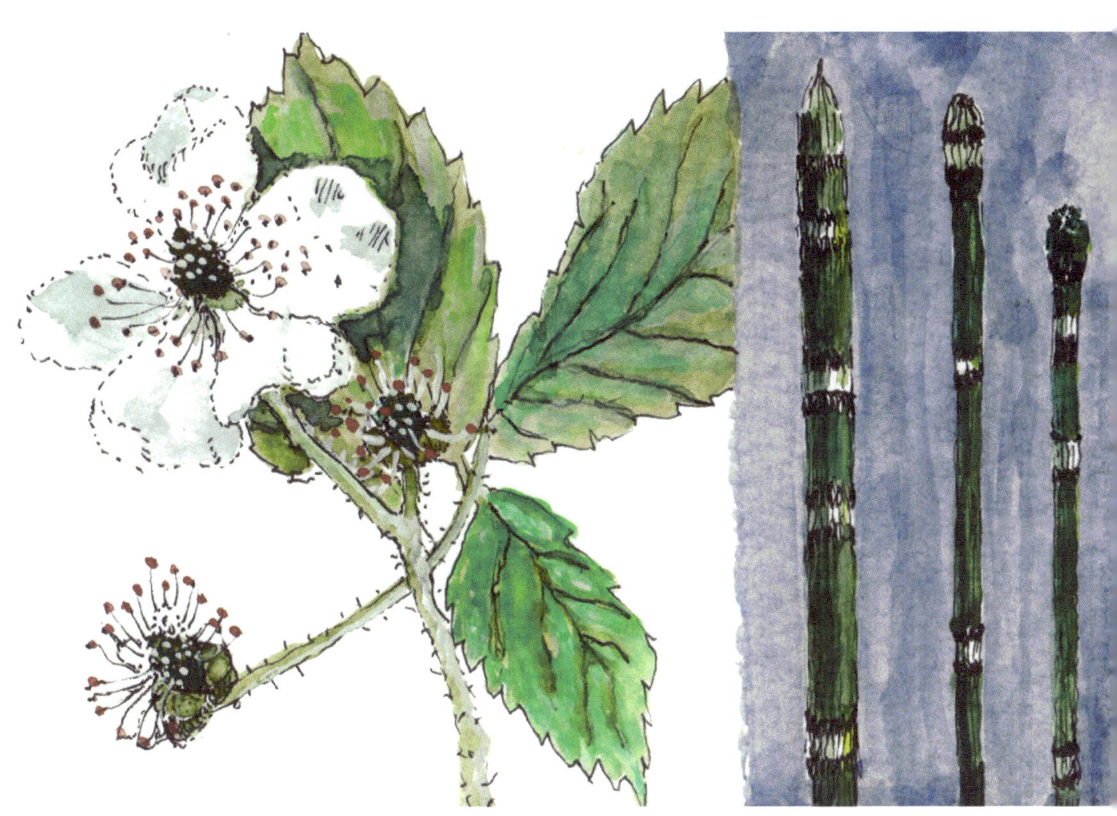

MOUNTAIN LAUREL

Lovely mountain laurel (kalmia latifolia in the *Ericaceae* or heather family) is blooming in sweet but poisonous abundance around here right now. This plant doesn't protect itself with thorns or stinging nettle-like whiskers. It seems to invite touching. One of the most astonishing and entertaining things about the flower is the way the stamens are each delicately tucked into little, red-marked dimples so that when an insect touches the bent stamen it springs up and shoots its seed off the flower.

Meanwhile, every part of this lush and aromatic plant is toxic to humans and some other mammals. Bees are its main pollinators, and the plant is not toxic to them, but the lethal-to-humans grayanotoxanes are carried to the hive and incorporated into the honey. Beekeepers need to keep their hives away from mountain laurel (as well as azaleas and rhododendrons) so that the honey from the hives does not hurt the people who buy local honey. Just handling the plant can give you a rash. Eating any part of it has been known to cause mild to severe gastric problems.

That said, mountain laurel has been used for centuries in a homeopathic way to treat skin and scalp problems among others. Native people know how to make a dye from it to dye fabric and hair. I was wondering how the plant manages to not poison itself, and then I read that the grayanotoxanes are sequestered inside special cells that move the poison around without harming the plant.

To the right is ground ivy or creeping Charlie, as ubiquitous around here as is English ivy. It was brought from Europe by early colonists as a beautiful cultivar. Happily, this member of the mint *(Lamiaceae)*

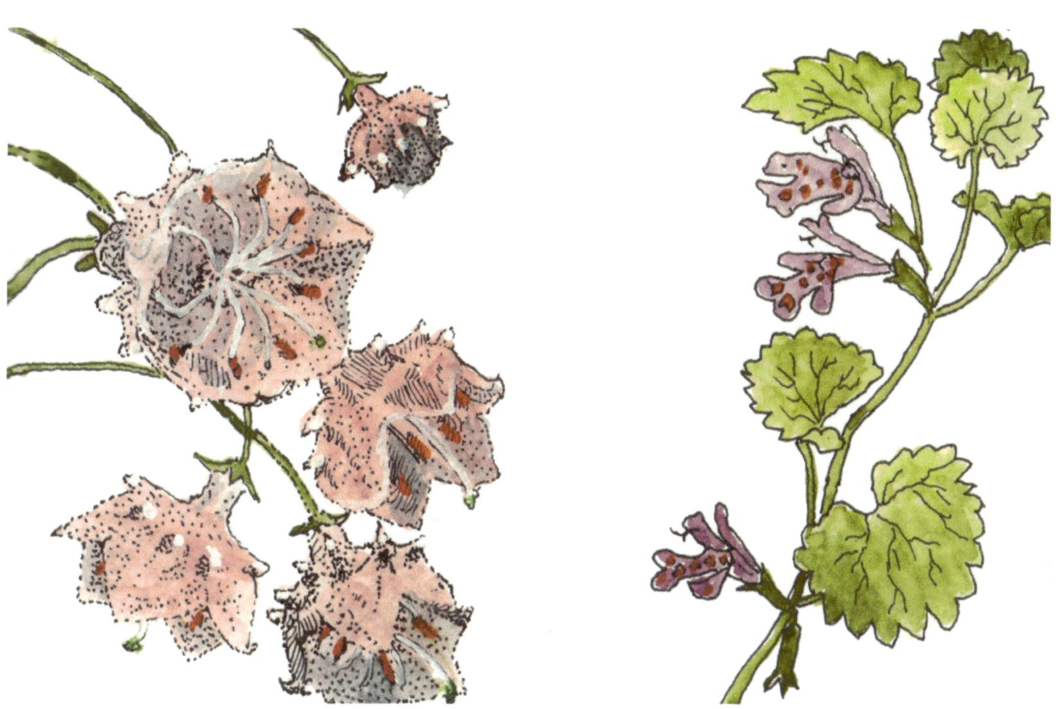

family is edible and makes a nice tisane that tastes minty and helps settle an unhappy tummy. The whole plant can be steamed and used like spinach or tossed raw into a salad. I would estimate that ground ivy covers 50% of our so-called lawn and is very pretty with its long-blooming tiny blue violet flowers. Above all, it never needs mowing.

https://en.wikipedia.org/wiki/Kalmia_latifolia

EGYPTIAN WALKING ONIONS

Egyptian Walking Onions (*Amarylladaceae* family) are on the move! Also known as tree onions, these natives of the Indian sub-continent form bulblets at the tops of their stems, as garlic plants do. As the baby bulbs mature, they get heavier and bend the long stems over, several feet from the mother onion. New plants can sprout from the bulblets; thus, your onions walk to new locations.

My babies left their mother's location this year and climbed into my square foot garden. The bulblets are delicious raw or cooked, and the stalks make good green onions. Best to leave the mother undisturbed as she is rather tough and, left alone, she will continue to sprout new bulblet blooms.

https://www.egyptianwalkingonion.com

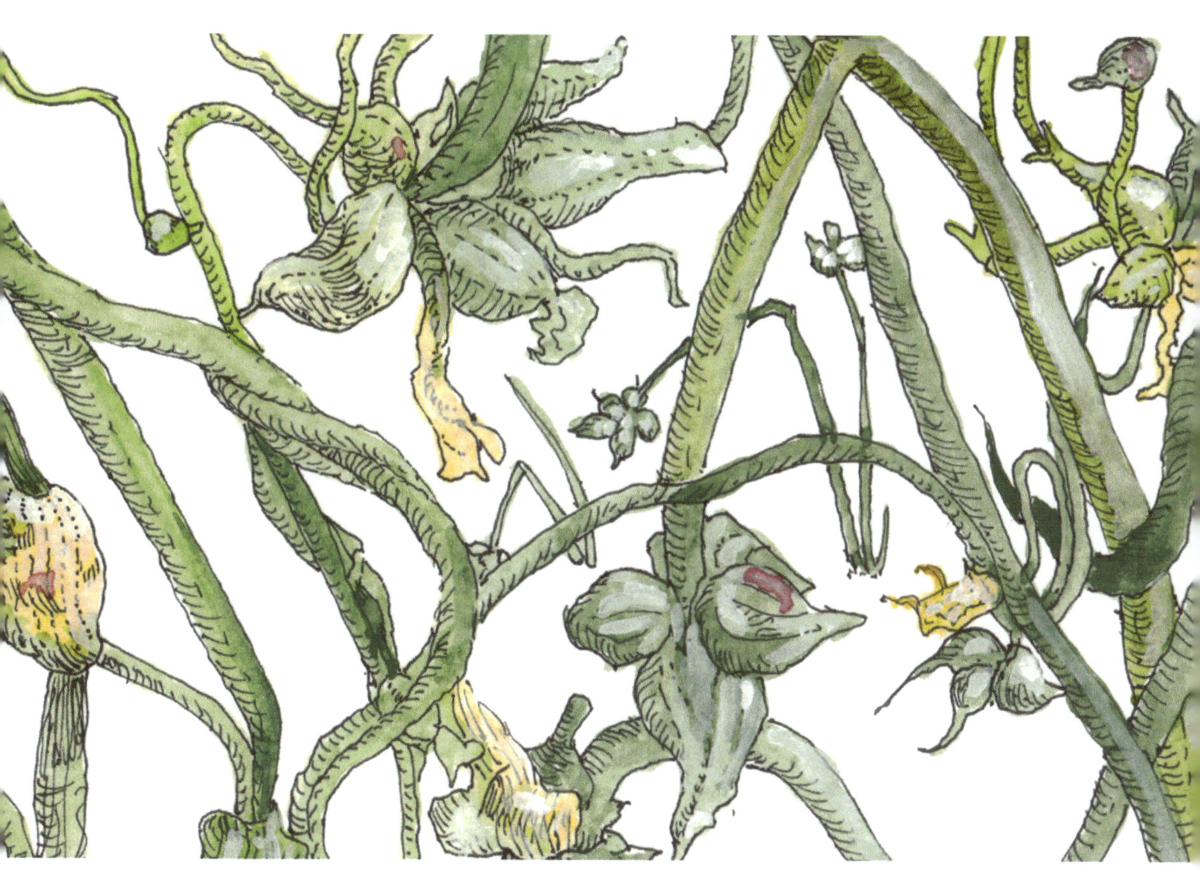

PEONY

The edge of the bouquet of peonies that is making our house smell wonderful today! I practically buried my nose in the blooms while I sketched the two buds. And peonies (*Paeoniaceae* family) are not poisonous to humans! Actually, they (especially the roots) have been and still are used in ancient medicines as well as in modern pharmacology. You can also parboil the petals when they drop and sweeten them with a little honey to nibble on with your afternoon tea.

Another fun fact – the ants that are always found on peonies are attracted to the sweet nectar that coats the blooms; they are harmless to the plant and are thought to help ward off predators. We brought some of our peonies from Michigan when we moved here in 1985. They have survived our three moves in the Asheville area, and one was accidentally buried under three feet of soil when we were having terraces made in our first garden here. We thought we had moved all the plants, but at the very end of the summer the lost peony made its way to the surface and bloomed late in the summer.

https://en.wikipedia.org/wiki/Peony

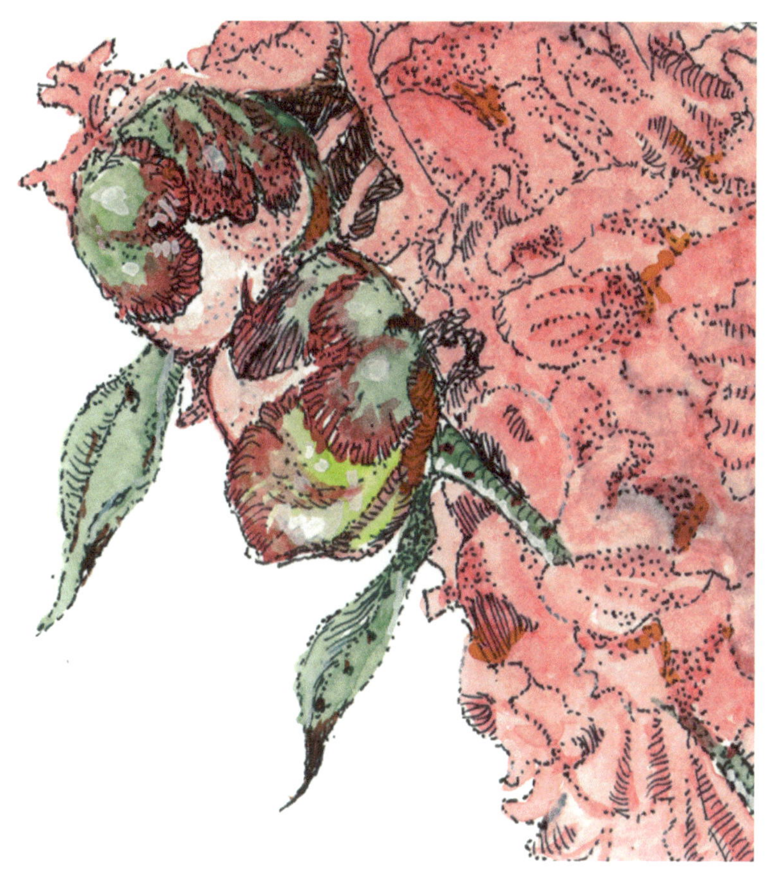

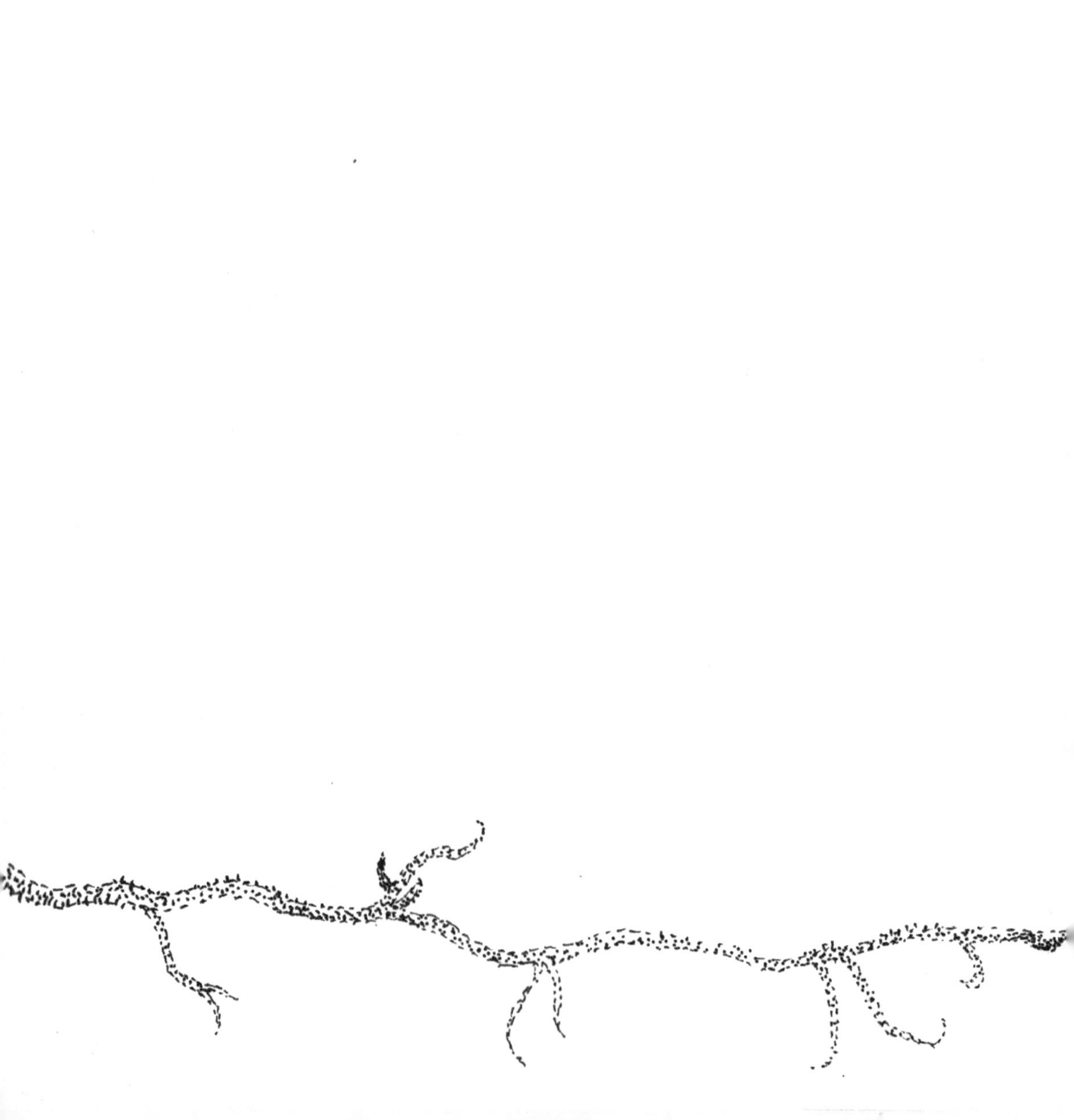

Where does the forsythia end
and the bee begin?

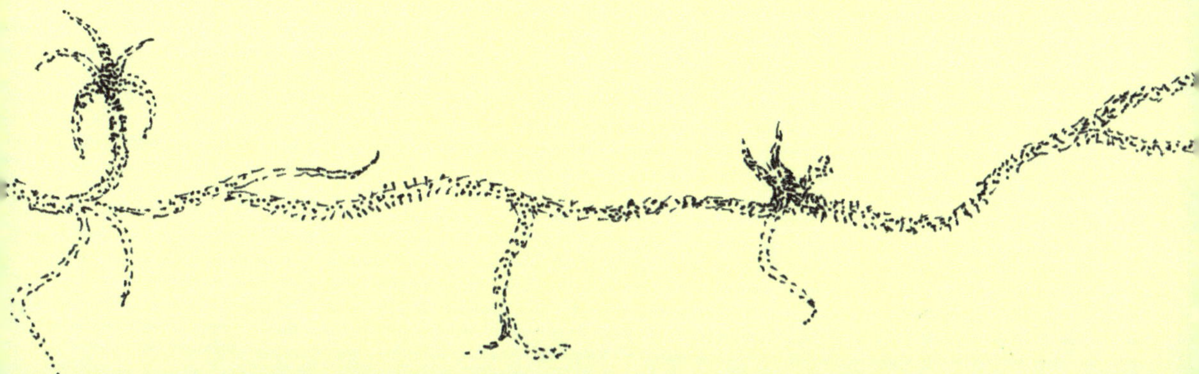

EGYPTIAN WALKING ONIONS -
bouquet

Bouquet: a few of last week's Egyptian Walking Onions have grown little bouquets on the top of their very long (23- 26") stalks! This one is in bud form still. Before it opened the bouquet was wrapped in a translucent membrane, which gave meaning to the old term for tracing paper: onion skin. You can see remnants of the membrane on both sides of the bouquet.

Green onions, which are the same thing as scallions, are excellent for your eyes, heart, etc., because of their supply of vitamin A and other vitamins, antioxidants, and other health-giving properties. A slice of an onion bulb will help stop the swelling and stinging of a bee sting. Here's a link to a delicious recipe featuring green onions:

https://healthynibblesandbits.com/ginger-shallot.../

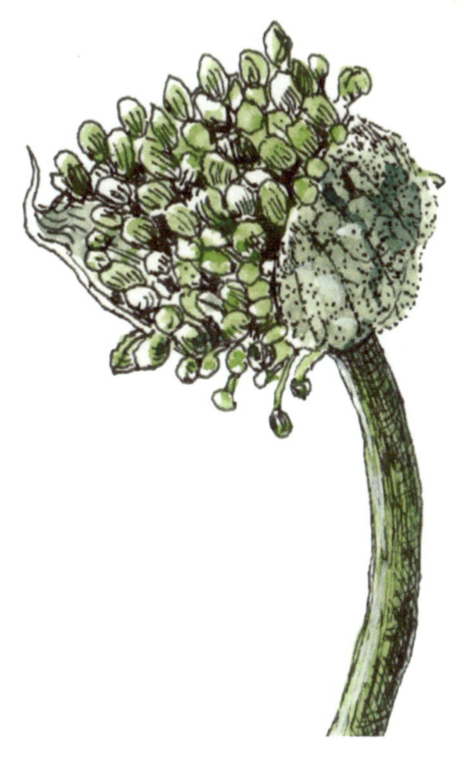

EGYPTIAN WALKING ONIONS -
knotting themselves

Walking onions blooming, tiny stalk-end bouquets flowering,
tying themselves in knots ...

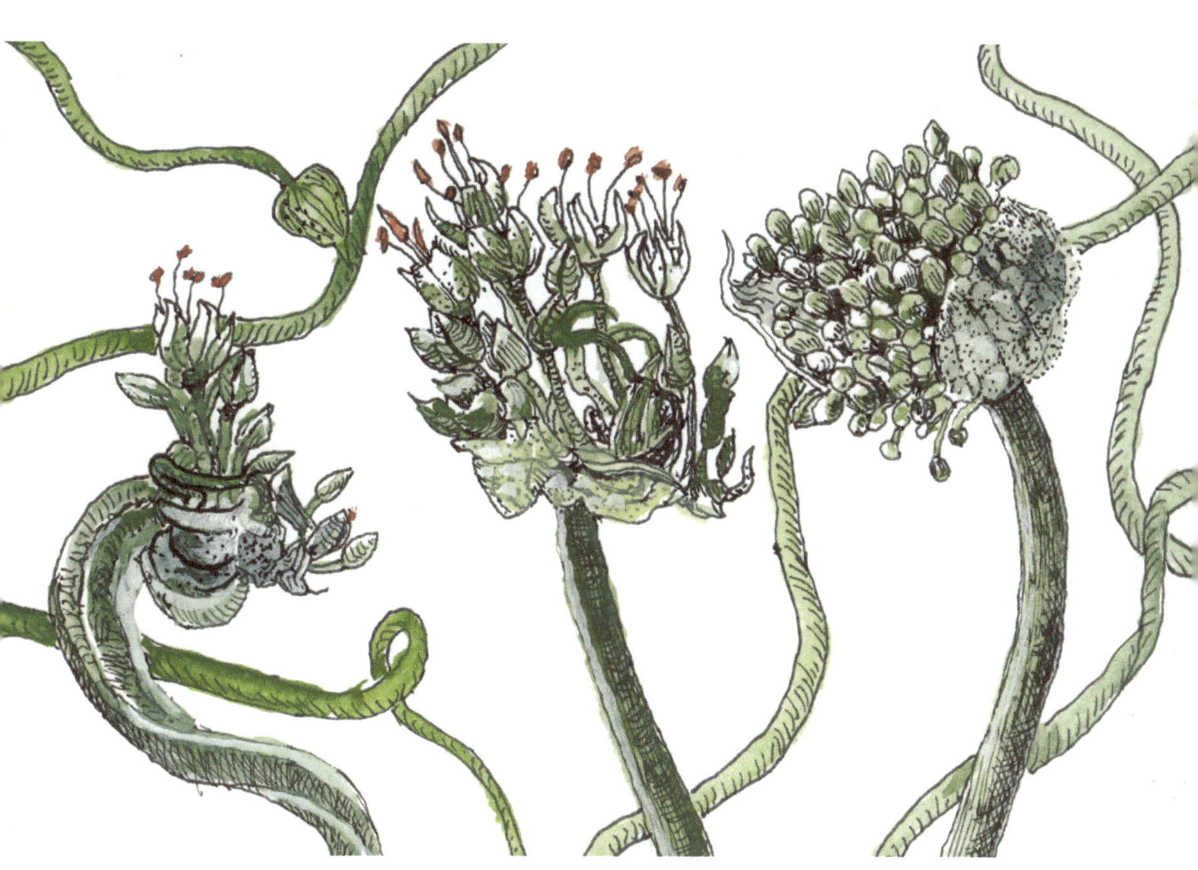

CLASPING BELLFLOWER

All along the river trail, flowers have faded, gone to seed, and been swallowed by the flourishing greens of leaves. But there are a few very small, unshowy blooms that now seem to pop out in the absence of all the yellows and pinks of spring. Here is the delicate Venus' Looking Glass or Clasping Bellflower, a member of the *Campanulaceae* family. It seems to have gotten its name from its shiny, reddish, mirror-like seeds that are enclosed in long capsules (none yet on this specimen).

The leaves are interesting in that they are clasping, cup-like sea-shell-shaped saucers. Only the top flowers open and are pollinated by bumble bees and potter bees; yet the bottom, unopened buds also produce seeds by means of self-pollination. So, the top flowers are chasmogamous, and the self-pollinating bottom buds are cleistogamous. (I knew that violets were able to cleistogamously produce seeds when they were in stress, such as during a very dry summer, but did not know other plants also did that. This plant apparently produces seeds both ways.)

Venus' Looking Glass leaves and blooms are edible, but they are so small and thin I would think you would add them to a salad or sandwich as though they were sprouts. The Cherokee made good use of the aerial parts and especially the roots by boiling them and using the infusion as a tisane or herbal tea for indigestion. They also added the tisane to bath water for the same purpose.

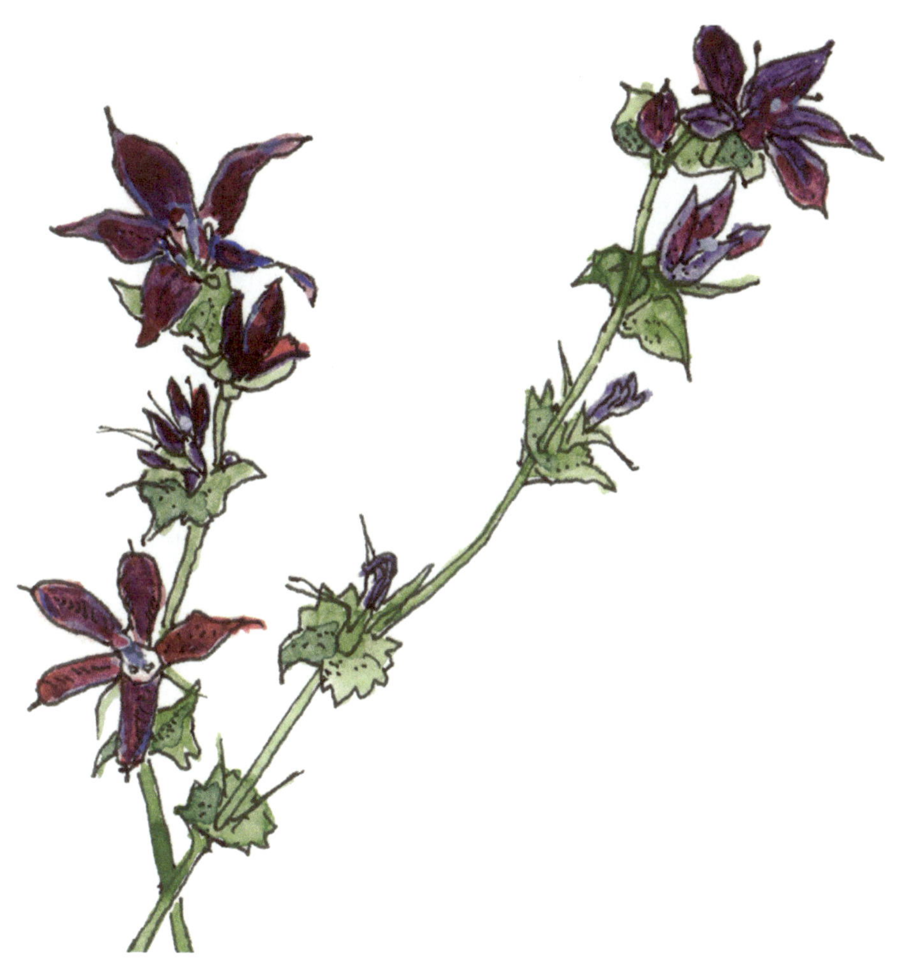

WILD GARLIC

Tops and bottoms of wild garlic (*Amarylladaceae* family), one of the most healthful, easy-to-forage plants around. They're easy to spot sticking up on thin, wiry stems with their purple-ruby colored bulblet clusters. There are literally hundreds of varieties, and you can know for sure that they are edible if the leaves, stems, flowers and bulblets have a garlicky smell.

Wild garlic has been found fossilized in thousands-of-years-old sites. It has been used as medicine as well as delicious food, excellent for your heart and to lower blood pressure. When you find the ruby-colored bulblets, you can gently tug the long stems to harvest the garlic. At the same time, you should rub the bulblets to free the tiny seed bulbs to scatter for next year's garlic. Then clean off the bulbous roots and hang them up to dry in a braid. They will supply you with delicious, healthful garlic for months.

https://www.thespruceeats.com/what-is-wild-garlic-435437

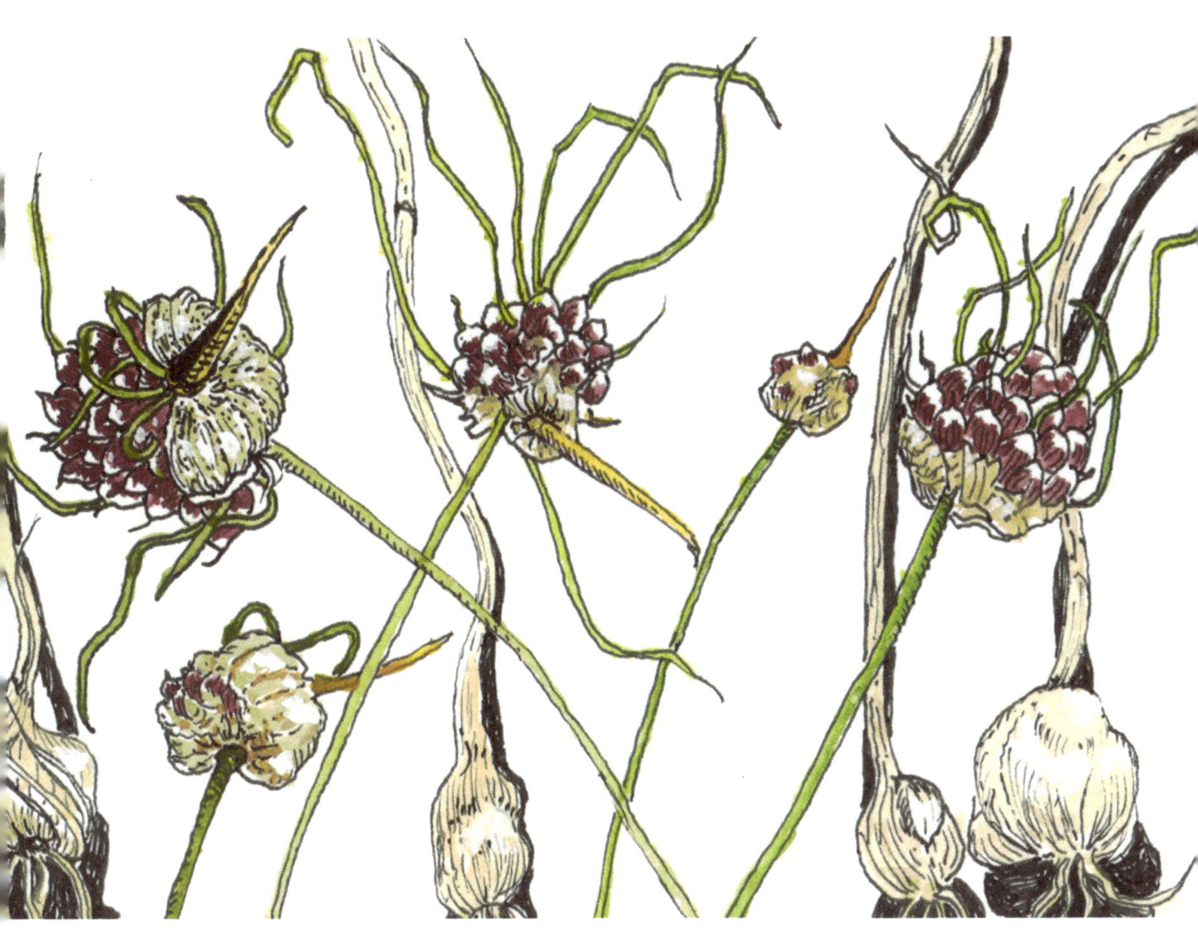

INDIAN PINK

Indian Pink (spigelia, *Loganiaceae* family) is such a glowing red and yellow flower that it's easy to see why hummingbirds love its tubular flower. I drew this in the UNCA botanical gardens last week. The dried root of Indian pink has been used by humans to get rid of tapeworms. In one article I learned that homeopathic spigelia has been useful in helping people get over a fear of pointed objects. Hmmm.

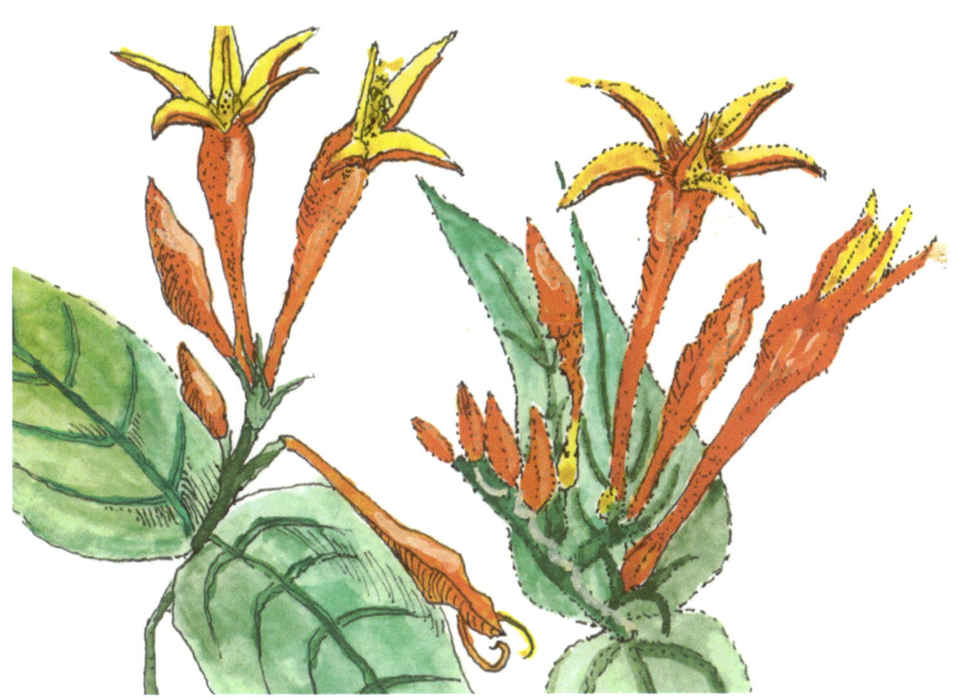

DAYLILIES

Daylilies, (*Asphodelaceae* family) like all plants, are a part of the interconnected web of plants-air-earth-animals. Ephemeral as their lovely blossoms are with their blooms that last for only twelve hours, the plants not only feed pollinators and produce seeds, but they alchemically clean the air that surrounds them. Many of them live in ditches that run alongside roads (one common name of the orange and yellow naturalized ones is ditch lilies), and they remove heavy metals from car exhaust as well as other toxins and carbon dioxide, transforming some of these into oxygen, without which humans and other animals could not live. And they do this mighty job while also brightening the earth and providing food and medicines for humans.

https://en.wikipedia.org/wiki/Hemerocallis_fulva

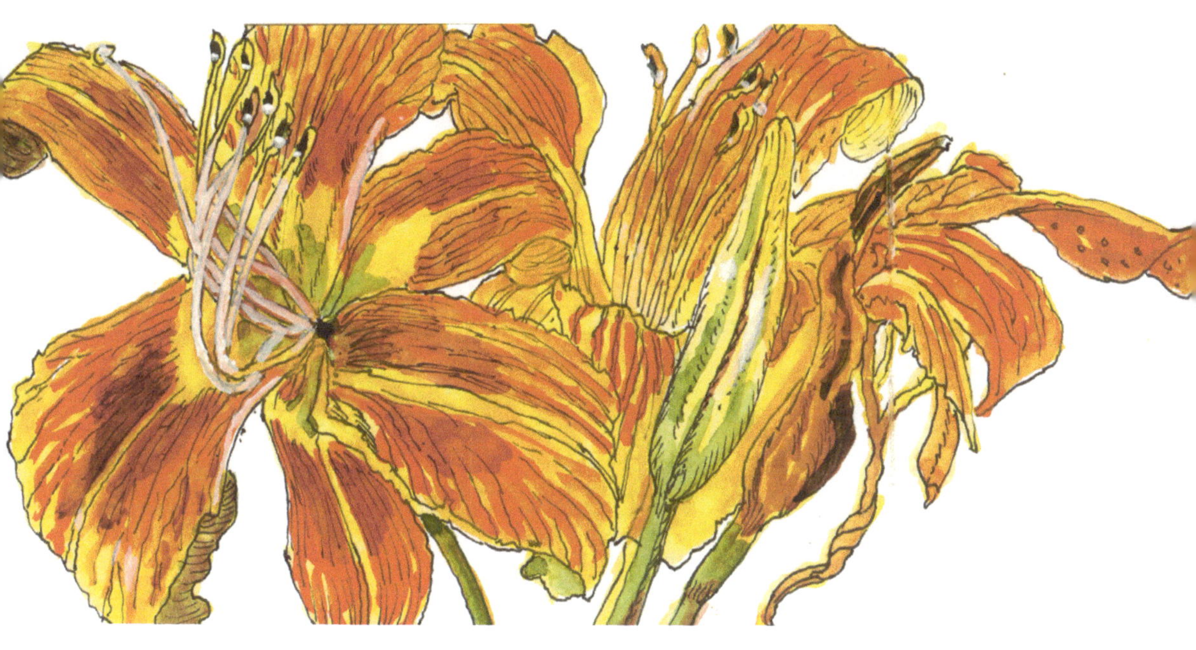

YELLOW CHESTNUT TREE -
buckeye survivor

Not as lovely as a daylily, but very exciting to me is the only buckeye to survive on the buckeye tree bloom I've been following for months. There were six baby buckeyes to start with, and this is the only one still surviving. Buckeye nuts are shiny and brown and are traditional good luck charms. The whole yellow chestnut tree that produces buckeyes is poisonous to humans and other animals. Only squirrels eat buckeyes, and even they seem to take only a few bites out of the nuts. Native Americans developed lengthy processes to leach out the poisonous tannins and saponins and render edible the protein-rich meat of the nut, which they ate and also used as medicine for pain.

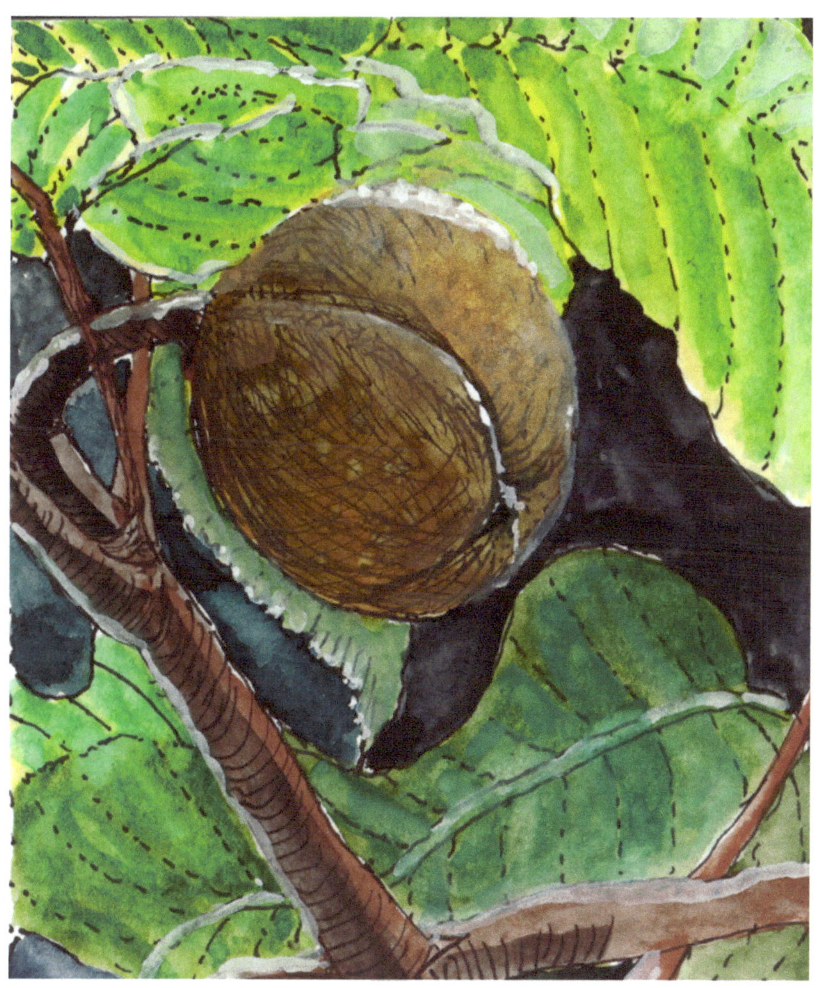

BEE BALM

Bee balm or bergamot is popping up all over! This member of the mint family *(Lameaceae)* is good for so many things in addition to its intricate architecture and subtle coloration. How could something that looks like a ridiculous 19th century hat also be an ancient medicine, a non-poisonous mosquito repellent, a delicious addition to salads as well as to pizza and bread sticks?

The square stem is the give-away that this is a member of the mint family. You might recognize the scent of the leaves and flowers as the flavor of Earl Grey tea. Native Americans, those clever and prescient people, have used it for centuries as an antibiotic to treat mouth and throat infections as well as minor cuts and wounds. Made into an essential oil, bergamot is useful for stress relief and pain relief as well as a mood elevator. (After leaning close to it in the UNC Botanical Garden last week for an hour while drawing it, I was in a decidedly happier mood than when I arrived at the garden.)

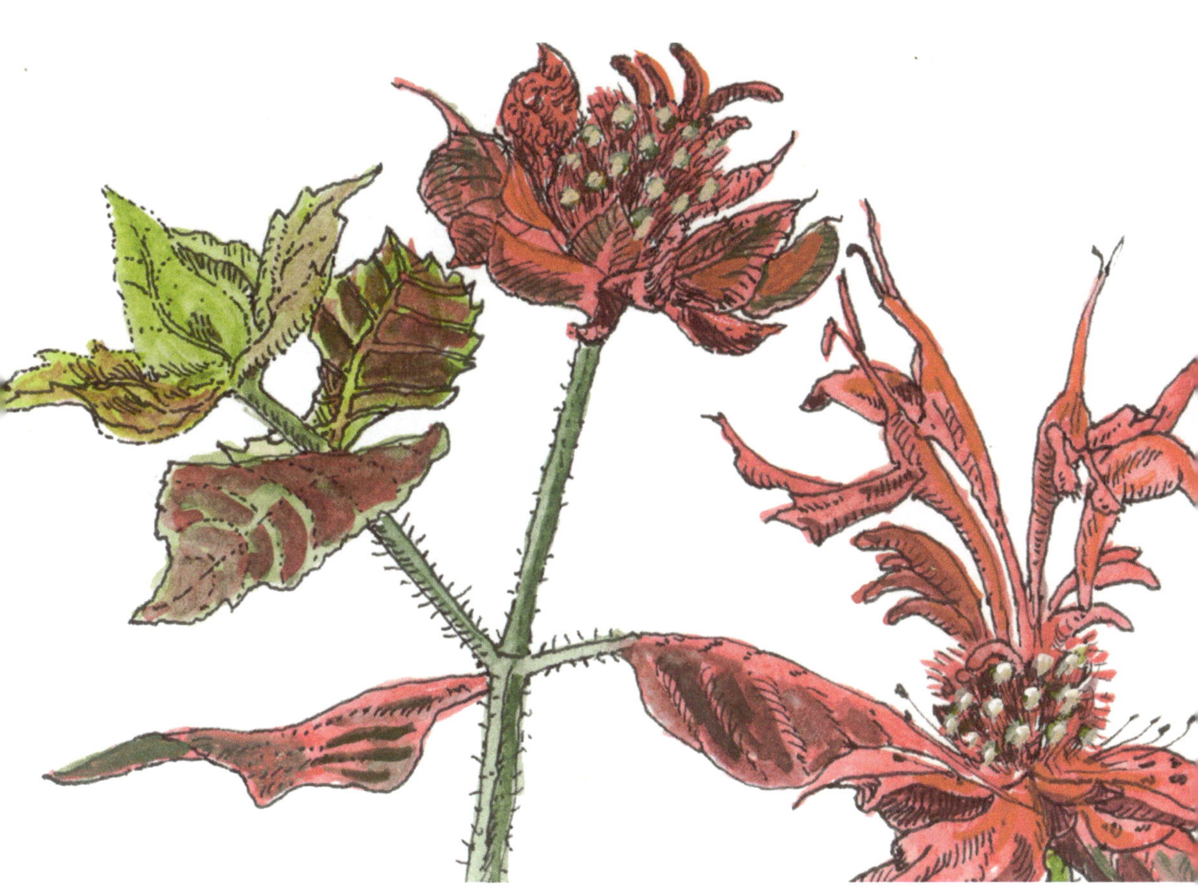

NASTURTIUMS

Nasturtiums and a few late honeysuckle flowers, what could be more cheering? (And today we DO need cheering up as we learned that we not only have masses of water under our kitchen floor but also asbestos in ancient linoleum mastic and possibly termites just to add to the fun.)

Back to the nasturtiums, though, as just sniffing their lusciously crimson and orange petals is a treat. Every part of this member of the *Tropaeolaceae* family is deliciously edible and packed with vitamin C, manganese, iron, flavonoids and beta carotene. This plant has been used medicinally for thousands of years as a cure for UTIs, coughs and colds, mild muscle aches, and scurvy. It is used to combat bacteria, viruses, and fungi.

Obviously nasturtium flowers are yummy in salads and on cakes, but how about in butter? You can chop the petals and leaves and knead them into softened butter for use on mashed potatoes, vegetables, and on bread.

These plants attract pollinators, repel many other insects, and even act as decoys to attract cabbage white butterflies away from cabbages, broccoli, and kale. It thrives in poor soil and tolerates neglect. Oh, and you can pickle its large seeds in vinegar to make capers.

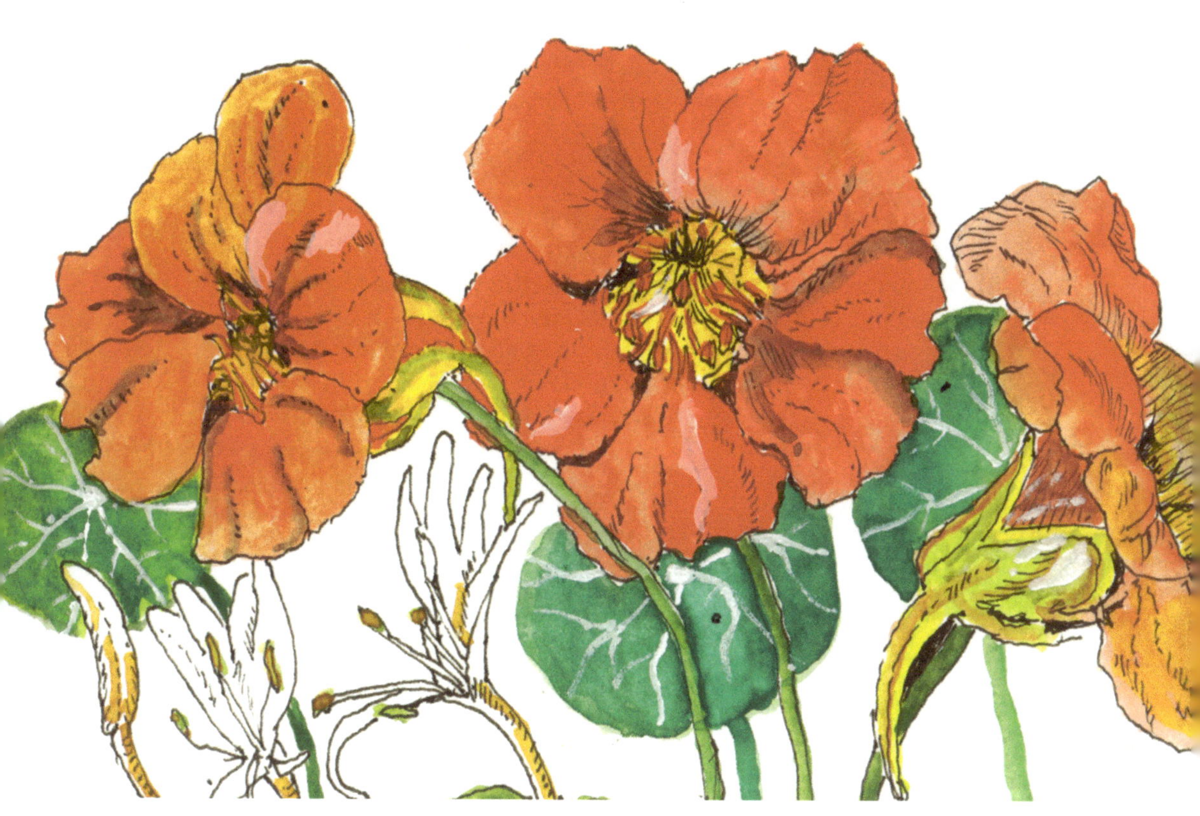

MAGNOLIA BLOOM

My son has a small magnolia tree in his yard, and he let me take one of its blooms yesterday. I feel like a time-traveler when I'm studying the tough carpels and leathery leaves: this plant has been around for more than 95,000,000 years. Imagine the world of the first *Magno-lioideae* family members – bees and butterflies had not yet evolved, and the early pollinators of these plants were relatively clumsy beetles; hence the tough flower parts necessary to withstand the rough pollination of the only pollinators available, beetles that were apparently after the protein-rich pollen of the magnolia flowers.

The bark and flowers of M. officianalis have been used in Traditional Chinese Medicine for thousands of years. I, personally, used to make fairy boats out of magnolia petals and pods when I was small child in New Orleans.

https://en.wikipedia.org/wiki/Magnolia

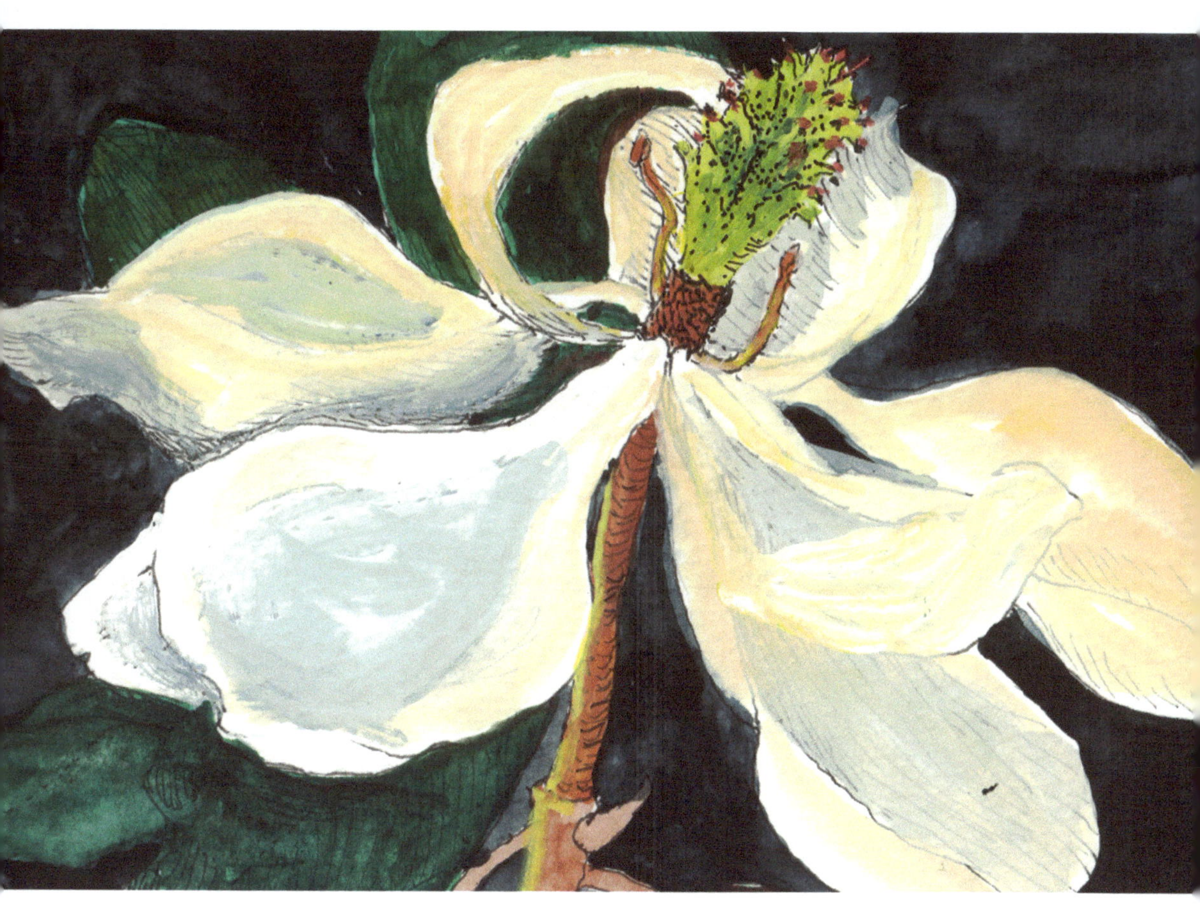

SELF-HEAL

I found a nice patch of self-heal, also known as heal-all, (Prunella vulgaris) at the south end water's edge of a pond in the park near the river trail. I've been searching for it all summer, but this is the first I've seen of it. This is a small, pretty, blue-flowered herb with the square stem characteristic of the mint family *(Lamiaceae)*. It has been used for centuries by humans as a medicinal herb for so many conditions that I'm posting the link to a site that will tell you more than you could imagine about this plant.

https://www.alchemy-works.com/prunella_vulgaris.html

Bees and butterflies love self heal too, and bunnies enjoy eating it. All parts of the plant are edible and beneficial. Make an infusion or tea with 1 tsp of dried herb in a cup of water just off the boil; steep ten minutes.

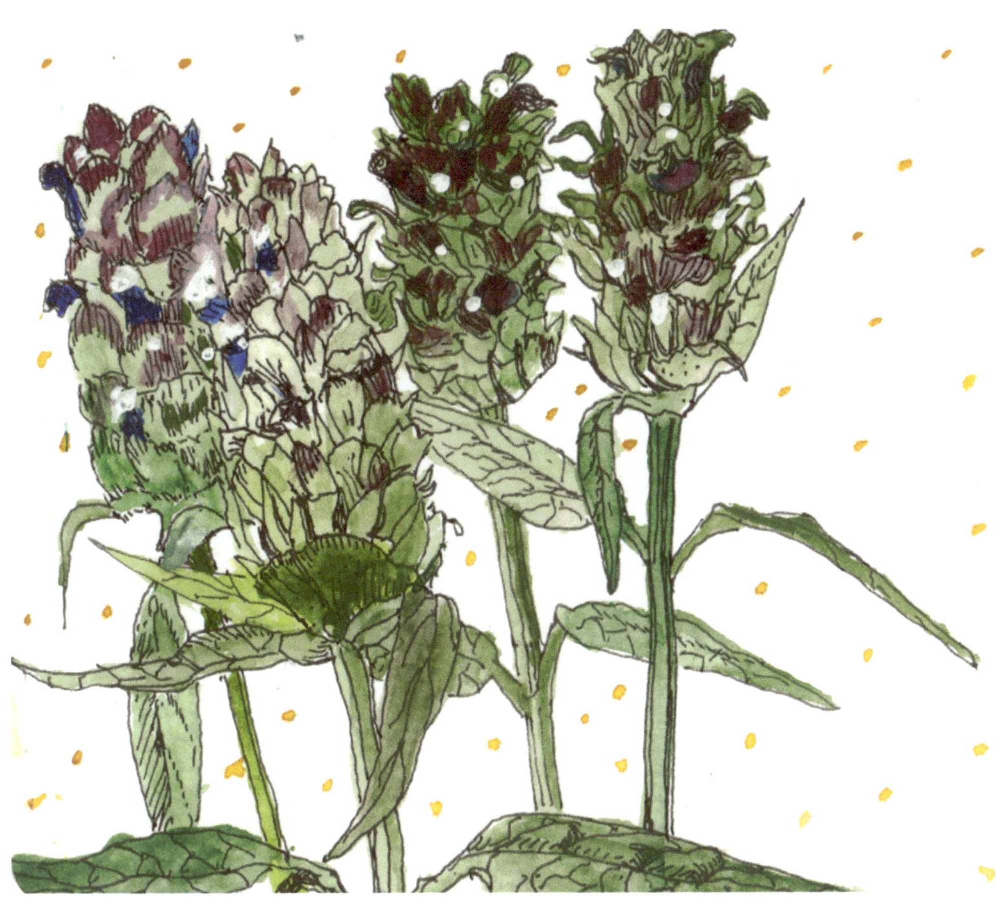

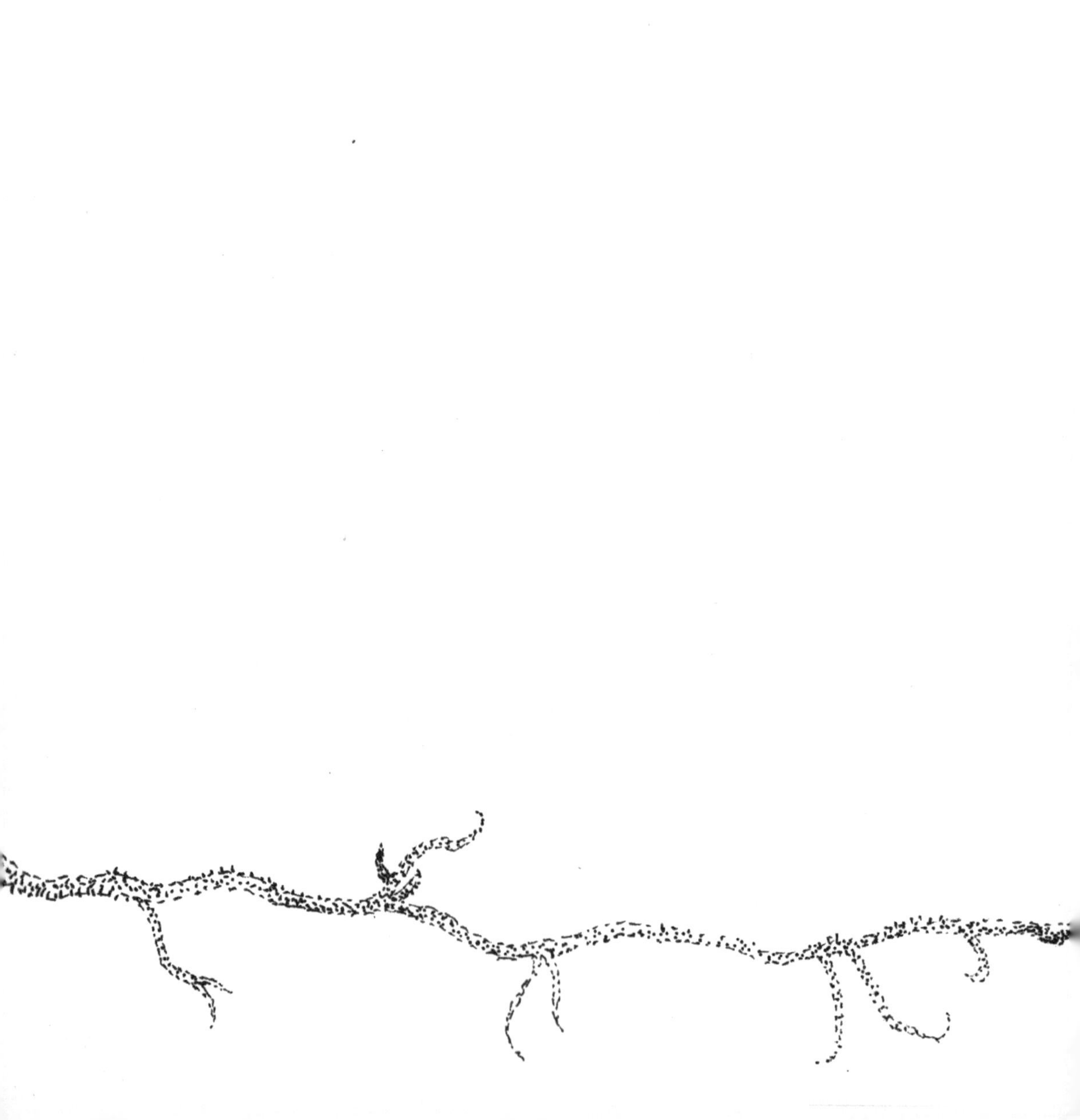

What is the thread that
ties the bee to the sun, to the
forsythia, to the honey, to the
human, and the bear?

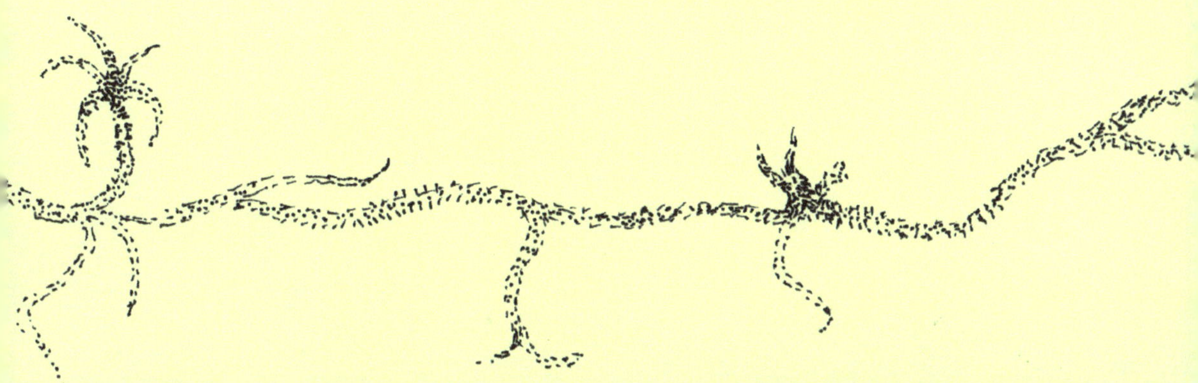

BEE BALM

The other day I came across a small patch of delicate white bee balm/monardia on the river trail. Monardia flowers look kind of frantic and messy in their mature bloom stage. But if you look straight down into their centers as they are just opening, before all the fire-crackery commotion, you can understand the geometry that underlies all of nature and informs so much of human design: the perfect Eastlake carved window corner perched up on top left of this drawing is as symmetrical as if it had been drawn with a compass and dividers.

https://www.pinterest.com/pin/155163149645862772/

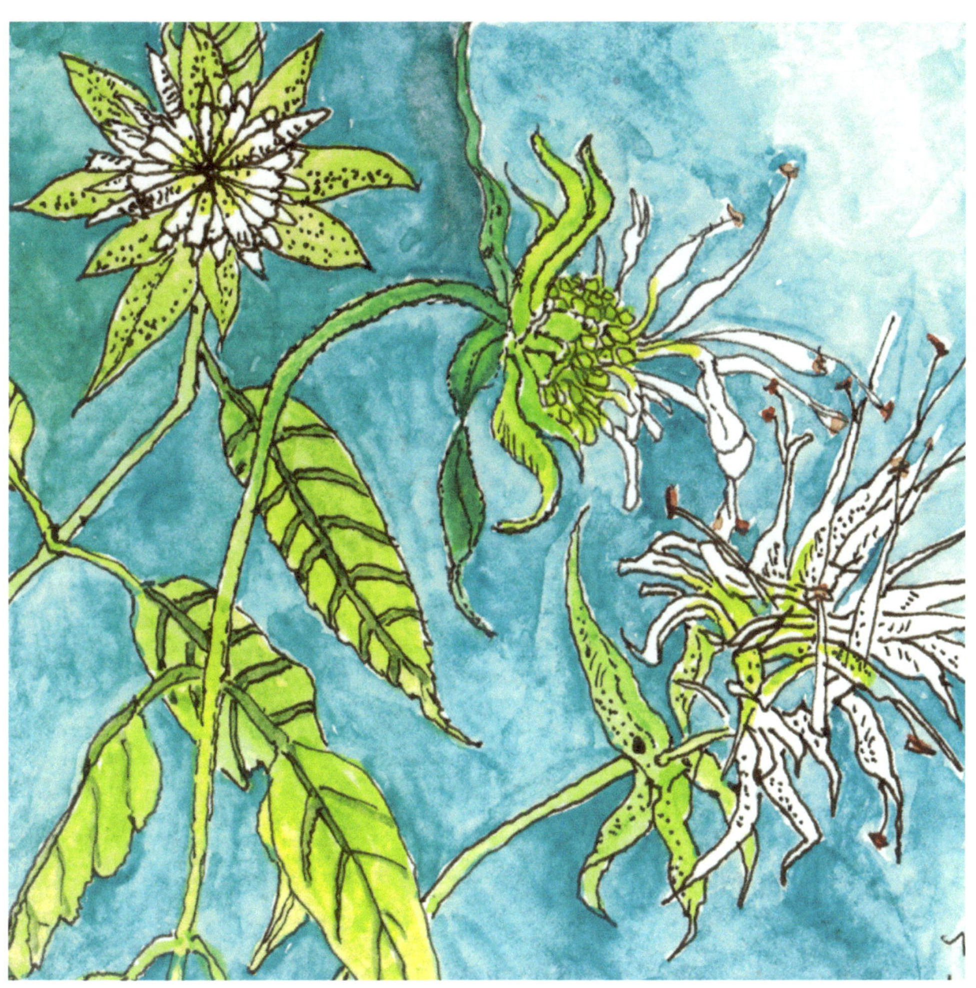

JEWEL WEED

Jewel Weed! I've been waiting for this member of the *Balsaminaceae* family to bloom all summer, wondering why it's so late this year. It's everywhere in our woods wherever poison ivy is (which is pretty much everywhere, even hanging in droopy garlands from trees).

Native people have always used the sap of this soft herbaceous plant to soothe the itch of poison ivy. So, when you're in the woods and accidentally brush against some of the dreaded shiny Leaves of Three, Let It Be, grab a big handful of jewel weed and rub its aloe-like sap on the spot. You can usually find it nearby, especially now that its bouncy orange and yellow flowers are busy attracting pollinating hummingbirds and bees and butterflies.

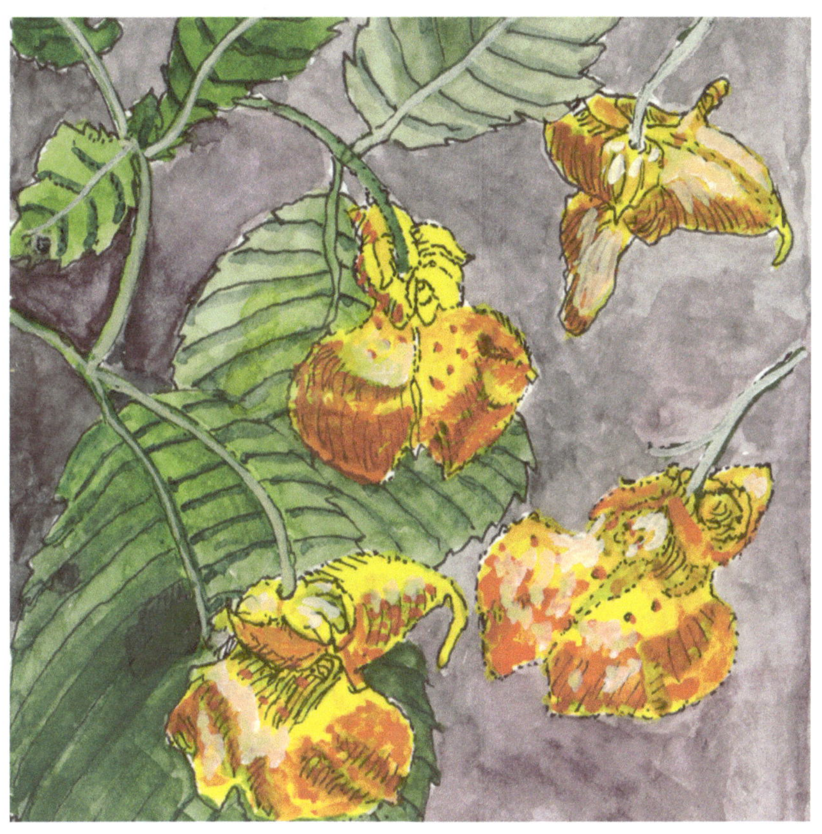

LUNARIA

As we move into Late Summer, there are signs of a shift from pollination and fruiting to seed production and releasing in many species. Today's drawing is of the seed stage of lunaria (*Brassicaceae* family): the astonishing money plant.

These are the now drying seed pods in their flat form, still attached to the plant. They look like silvery moons, hence the name; but astonishing as these are, there are even more delicious developments. As the pods mature, they dry out and then begin to peel apart as the dehiscent edge seam deteriorates and splits open. When the time is right, the top and bottom disks peel off, the flat blackish-brown seeds float out, and the silvery center disk is left on the plant with only the tiny umbilical stems left behind on them to mark where the seeds were developing.

Lunaria has been used in Traditional Chinese Medicine as well as in Native Medicine for thousands of years. In addition, the entire plant is edible and lends its mustardy-cabbage zing to salads and green vegetable dishes. The flat seeds can be mashed with water and let sit for 10-15 minutes to make a wild mustard.

https://en.m.wikipedia.org/wiki/Lunaria_annua

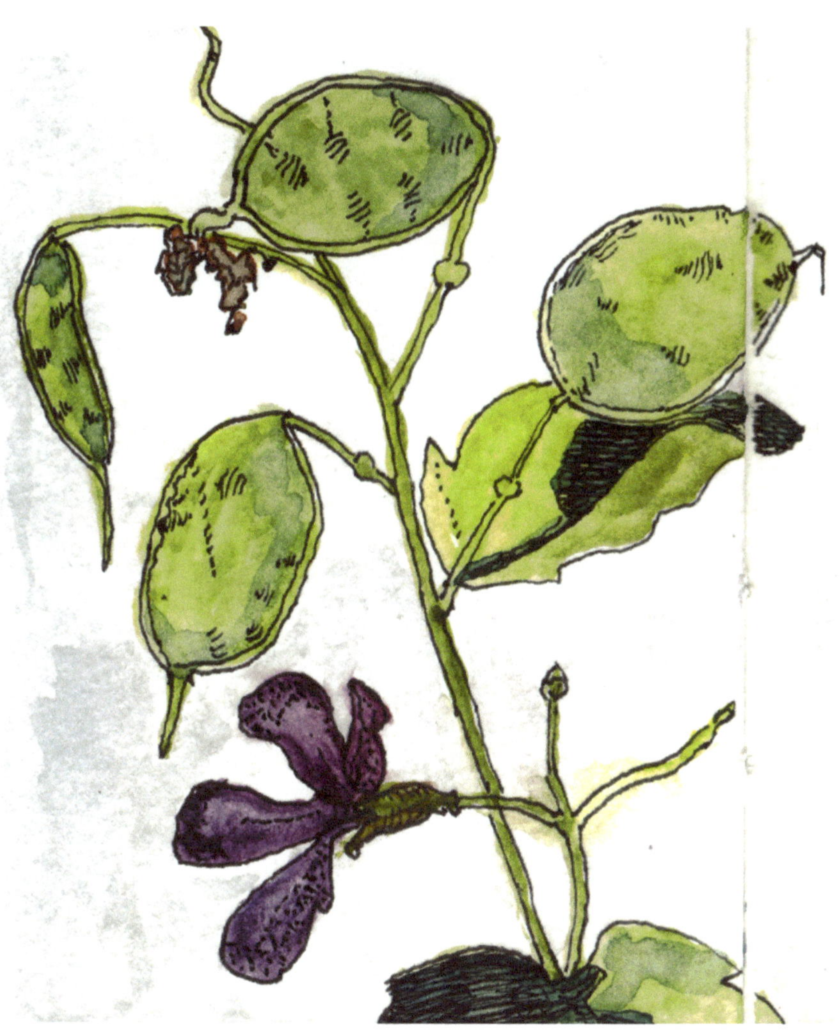

SOLOMON'S SEAL

Solomon's Seal (*Asparagaceae* family) is busy putting seeds into little dark blue fruits that dangle down beneath its frond-like leaf stems. The seeds are poisonous to humans but robins, bluebirds, veery, and wood thrush are among the many birds that feast on them and thereby help spread the seeds. The part of Solomon's Seal that humans have used for thousands of years is the root from which it gets its odd name. Native people in the Americas applied the powdered root directly to the skin for bruises, ulcers, boils, hemorrhoids, rashes, and edema.

Traditional Chinese Medicine uses the roots (called huang jin) in formulas for diabetes as well as fatigue and anorexia. It is also used to close wounds, reduce swelling, and reduce inflammation. The roots and young shoots are edible, used as flour and eaten like potatoes, and the shoots added to dishes of simmered greens.

The name of the plant derives from the scars left on the rhizome by old shoots. These scars were thought to resemble the seal of King Solomon, a ring that supposedly gave the king power over spirits and demons and that allowed him to speak with animals. There are many extant examples of the ring. In legends the design was engraved by God and given to the king directly from heaven. The two inverted triangles seem to have come out of a medieval Arabic tradition.

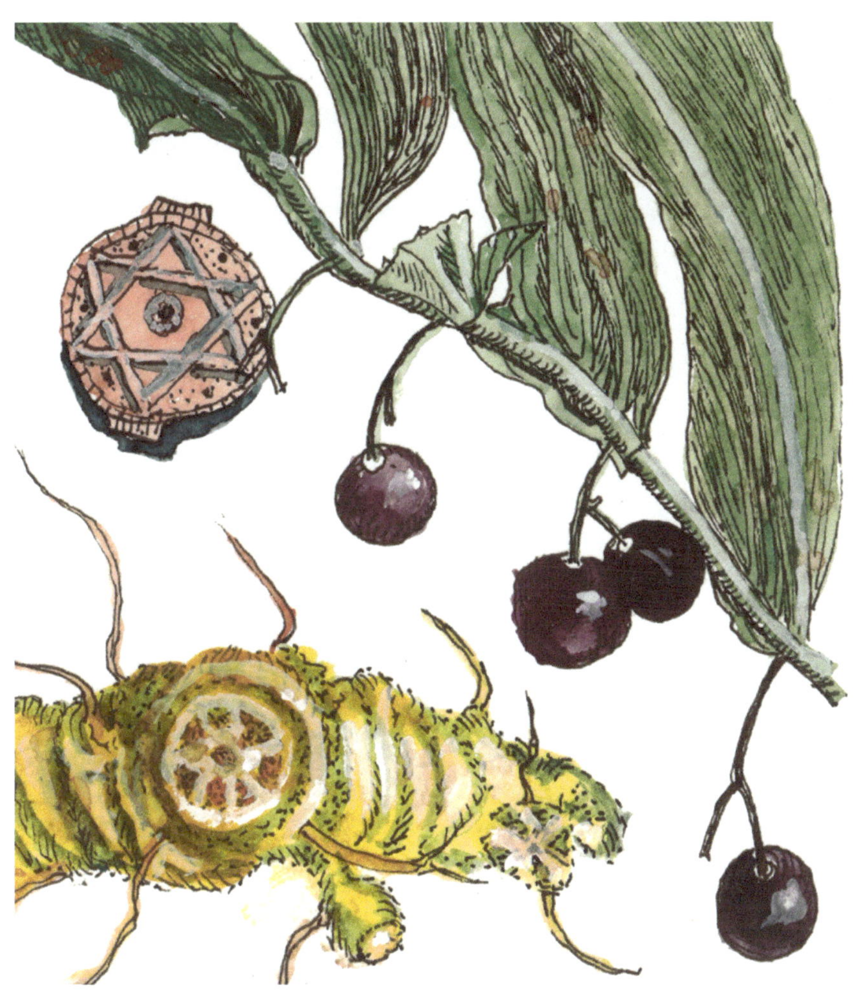

NUTS

Nuts are beginning to show up on the forest floor! Here are some black walnuts and pignut hickory nuts that I picked up this weekend. The big brown walnut in the center is an old one that is naturally degrading after over-wintering in the yard adjacent to a big antique store in Asheville. The dark brown pigment in the rotting exocarp is a mixture of tannin and juglone, and along with helping the seed or nut escape from its outer hull so that it can germinate, these pigments have been used for at least 4000 years by humans as a dye for hair as well as for protein fibers (wool and silk). Walnuts have also been a valuable food source for humans. Walnut meat has the highest protein level of any tree nut, as well as antioxidants and other health-giving compounds.

The walnut with the light green exocarp is not quite ripe and is from this year's crop. The smaller brown nut above the large green walnut is a walnut hull that has escaped from its exocarp but is still in its tanninous hull. The hull is extremely hard and difficult to remove. Years ago, a friend gave me a large bag of black walnuts and told me the best way to open them was to drive over them with a car.

On the left are three pignut hickory nuts, which serve as a good mast crop for bears, hogs, foxes, squirrels, deer, and other mammals. These nuts are also edible for humans, and they're mature when their husks turn from green to brown. The pignut hickory husks when ripe split open along dehiscent (or weakened) ridges.

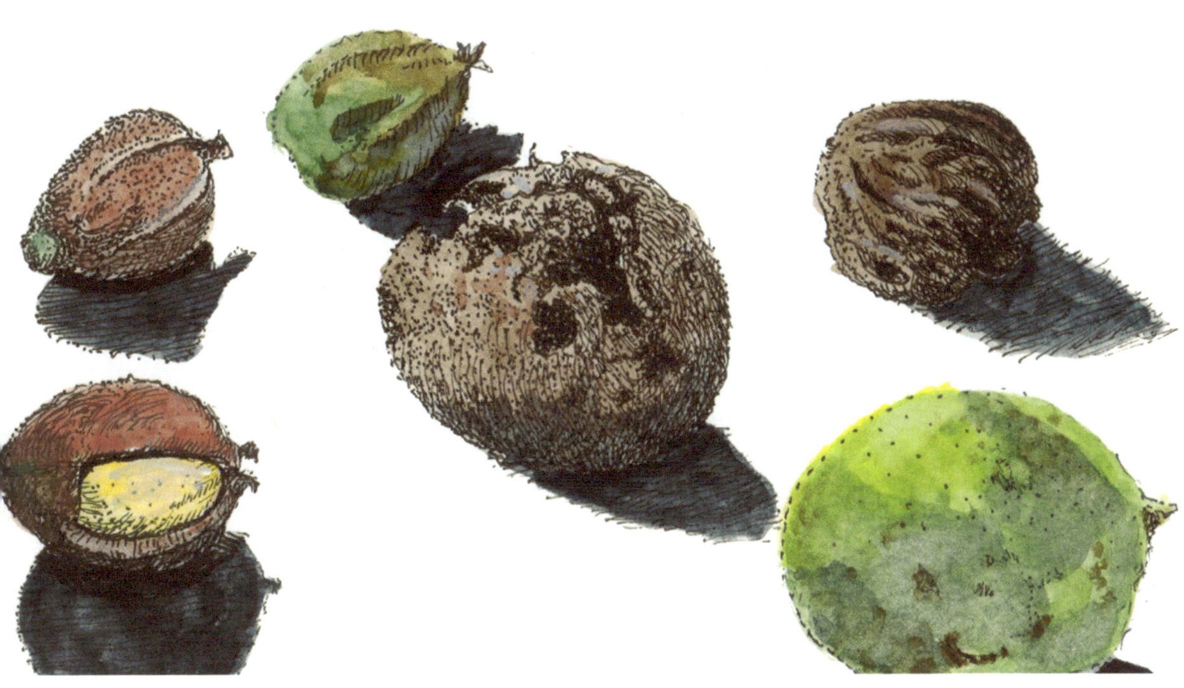

RED SOAPWORT

Red soapwort (Saponaria officinalis in the *Caryophyllaceae* family) has long been a source of gentle soap used especially to clean wool as well as delicate textiles. Legend has it that it has been used to treat the Shroud of Turin, and in Medieval times monks thought it had been given to humanity by God to keep themselves clean. Even today the roots and leaves are chopped and then boiled to extract the saponins for use in herbal shampoos and soaps. The liquid is said to be especially gentle as a soap replacement for dry skin.

In a more medical application, the chopped and boiled roots and leaves themselves can be directly applied to boils, psoriasis, eczema, and poison ivy. I found mention that the liquid has been used to heal bronchitis, coughs, and inflamed mucus membranes of the upper respiratory tract. The liquid is also used in beer production to help develop a good head and as an emulsifier in the production of tahini. Be aware, however, that ingesting too much of the liquid can cause nausea, vomiting, and diarrhea.

Red soapwort is pollinated by bumblebees and, especially at night, by moths, including the white-lined sphinx moth.

The red berries I could not identify. They look like they have been dipped in sparkle nail polish. I found them near the pond on a low bush. A few days later, my friend Michelle identified them as autumn olive, and I began to see them everywhere.

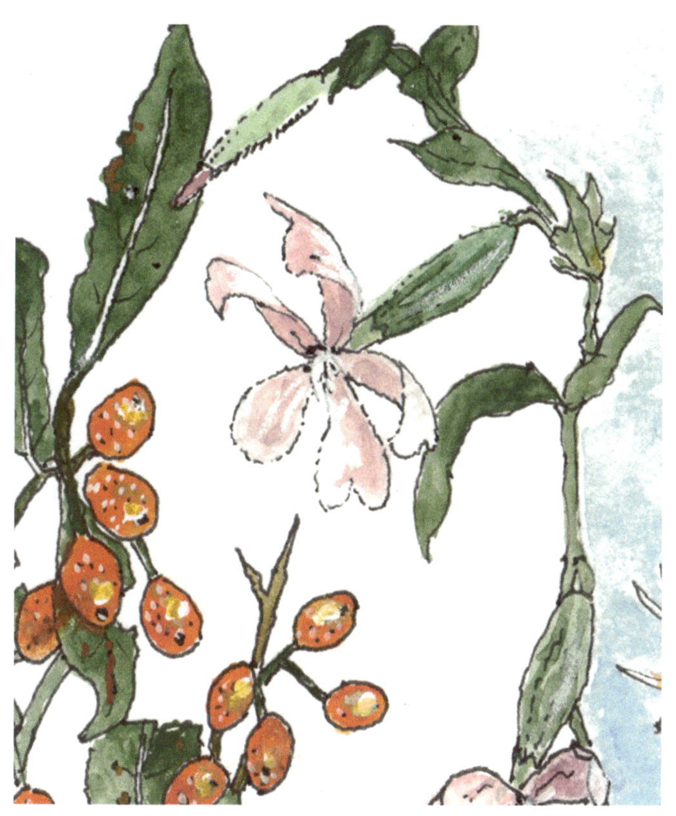

INDIAN PIPES

Deep in the old pine plantation on a hill at the back of campus some friends and I found numerous patches of the rare and mysterious Indian pipe plant. These seem to glow in the deep shade like little gatherings of ghosts. Totally lacking chlorophyll, these plants get their food through a saprophytic relationship with both russula mushrooms and trees, in this case pine trees. All the action takes place underground in the entangled root and mycelium systems, where the mushroom siphons off some of the tree's sugars in exchange for enhancing the tree's ability to absorb water and minerals. Meanwhile the Indian pipe absorbs carbon and energy from the mushroom, having changed its own roots so that the mushroom recognizes the Indian pipe roots as its own.

 The Indian pipe is an ancient plant with marvelously strange properties: it flowers in a bent over pipe-like position, hence its common name; if picked, the plant melts like warm gelatin and loses its form. When its seeds are ready to disperse, the flower turns upward and shoots out the seeds.

The plant has long been important in Native peoples' medicine as well as European homeopathy. The root is dug during August and September and carefully dried and saved to use to treat both physical and emotional pain. The plant is also a shamanistic plant in Native traditions.

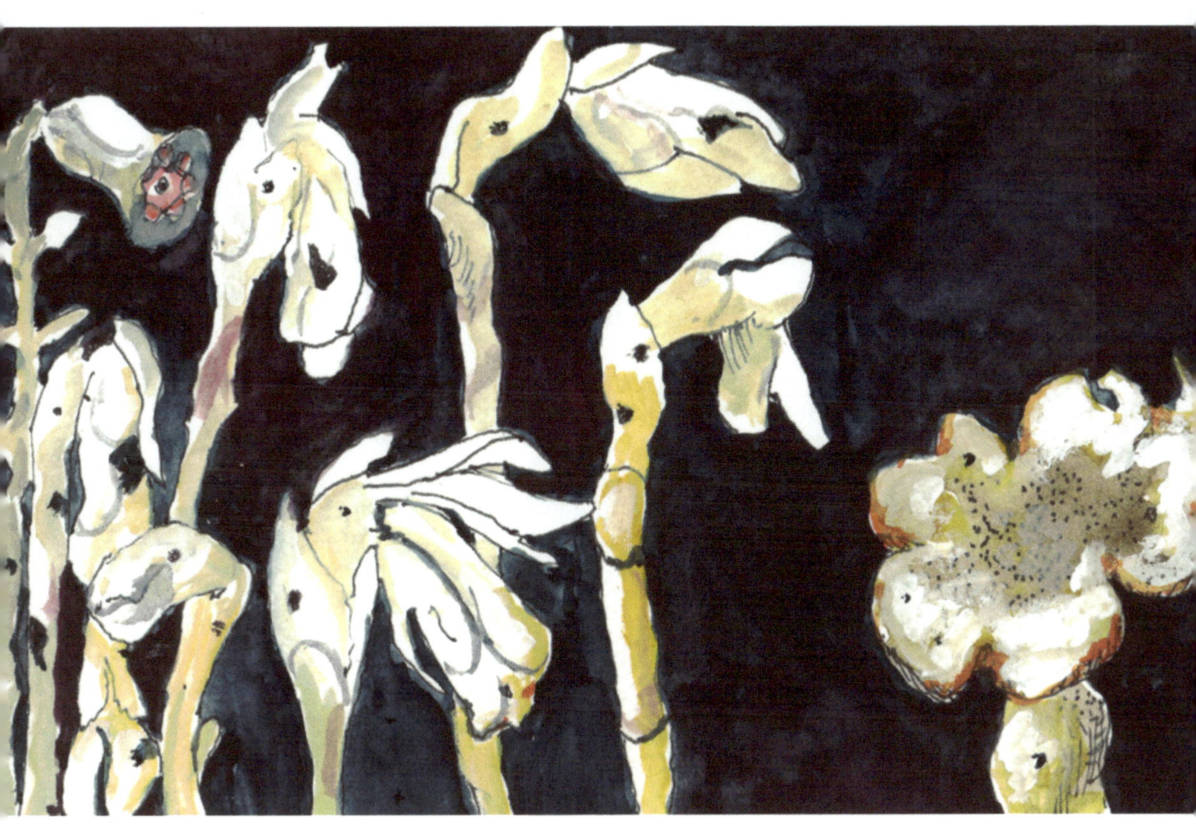

CORN

All spring and summer my granddaughter and I have been nurturing a large flower and Three Sisters Cherokee Garden that we planted in a bed that her brother dug and mulched with aged compost from our yard. It's finally producing actual ears of corn, squashes, and black beans along with sunflowers, zinnias, nasturtiums, pot marigolds.

We were planning on picking corn next week. But early this morning a mamma bear and her small babies visited and had a picnic of our corn. I managed to salvage a couple of ears that they missed and to stand up a few of the other corn plants that were just beginning to form ears. It was interesting to find the discarded eaten cobs, which looked like humans had eaten them. Our neighbor saw them heading down the street. Peeping into the frame on the left is a tiny wild orchid plant.

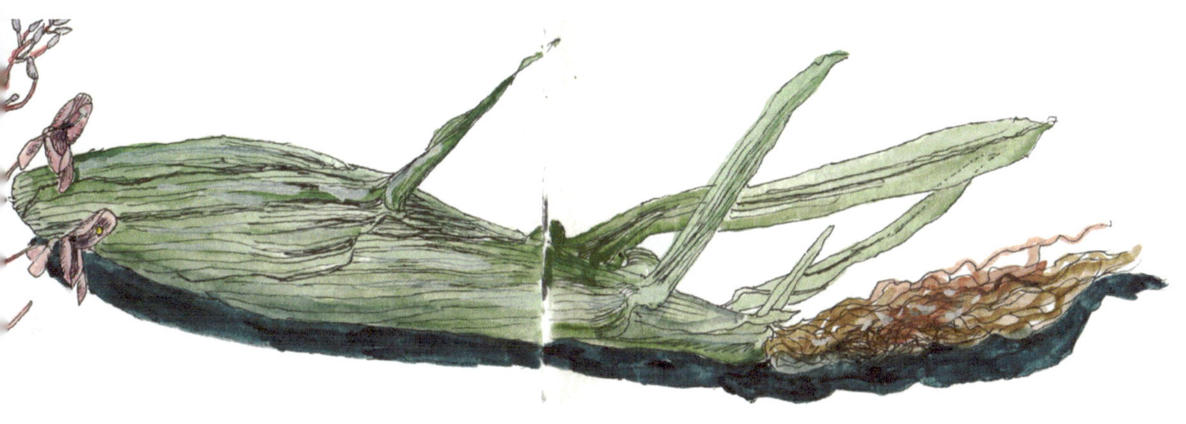

INDIGO PODS

These are indigo pods and leaves from the downtown park in Black Mountain. The luscious inky blackish blue pods give a hint about the pigment that might be in the plant, but in fact the green leaves are the source of the blue-black dye and pigment. Indigo is one of the few blues that exist in nature, but the dye itself doesn't exist in nature. In order to extract the indican, a precursor to indigo, the leaves must be soaked and fermented for 14 hours in a vat with the enzyme indimulsin added in order to break down the indican into indoxyl and glucose as well as to expel carbon dioxide.

The leaves contain only 2-4% indican, so many leaves must be harvested. After the fermentation, the indoxyl is mixed with air to make indigotin, which is then heated to stop fermentation. The resulting deep blue pigment settles to the bottom of the vat and is dried to a thick paste.

It's amazing to me that indigo dye has been made and used from as long ago as 3300 - 1300 BCE in the Indus Valley. How did people figure it out? Fragments of indigo-dyed cloth have been found in Egypt dating from 1600 BCE. Medieval Japanese warriors wore indigo-dyed garments under their armor for its medicinal qualities of healing cuts and bruises. Indigo was very expensive, as you might well imagine. At the top of the drawing is a triangle of indigo oil paint, the most transparent deep greenish black of blues.

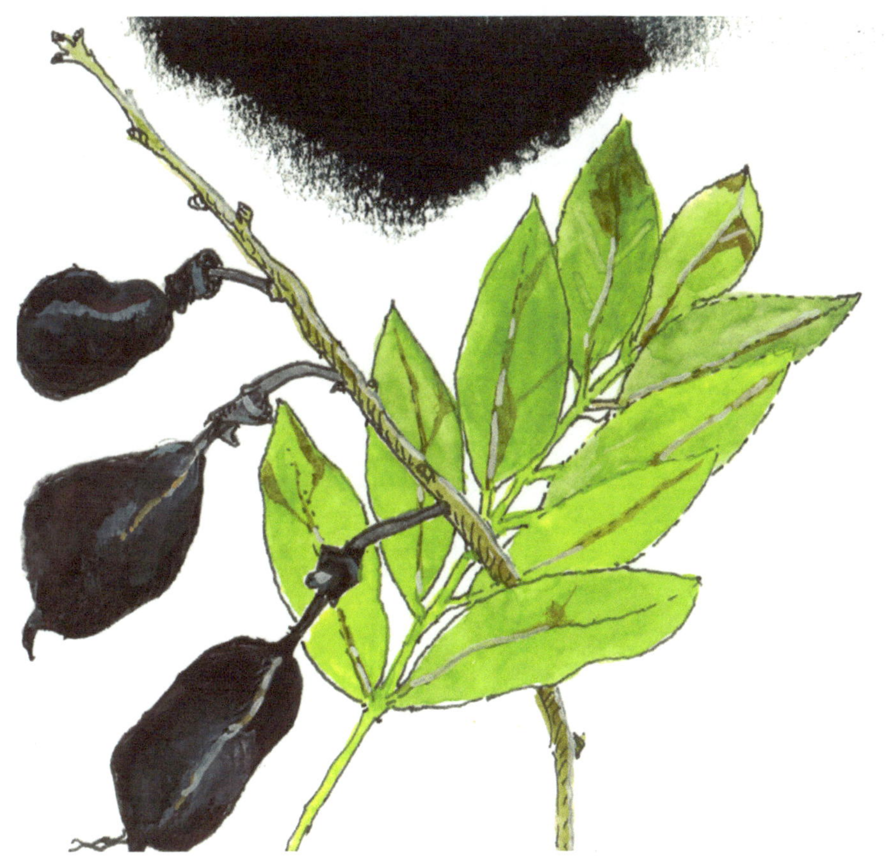

PODS, PODS, AND MORE PODS

Of course, the whole point of flowers is not to please humans with their luscious colors and fragrances, but to do their essential job of luring pollinators in to get the seed-making process going. And here are three different kinds of seeds still resting in their pods: one of the black bean pods plus a loose seed or bean, a popped-open indigo pod, and a barely open flag iris pod.

The pod or other structure that the plant has constructed to temporarily house the seed is designed to hold the seed in a dormant state until the correct amounts of moisture and heat cue the seed to sprout. Each plant has its own perfect way of guaranteeing that its seeds will not be released until conditions are just right.

In the case of the black bean, the pod must dry up and develop the brown tannins that make the pod bitter and unattractive to animals (including people). The dehiscent seam that runs along each pod ripens or degrades and releases in concert with development of the beans' cotyledons and the readiness of the germ to sprout.

The tough leathery pod of the indigo also has dehiscent seams (which are weaker places that split open sooner than the rest of the pod) and these degrade in concert with the maturing of the seeds. The round seeds themselves have tough seed coats that degrade after winter dormancy due to moisture and warmth.

The iris pod has wing-like structures from which the seeds grow. This

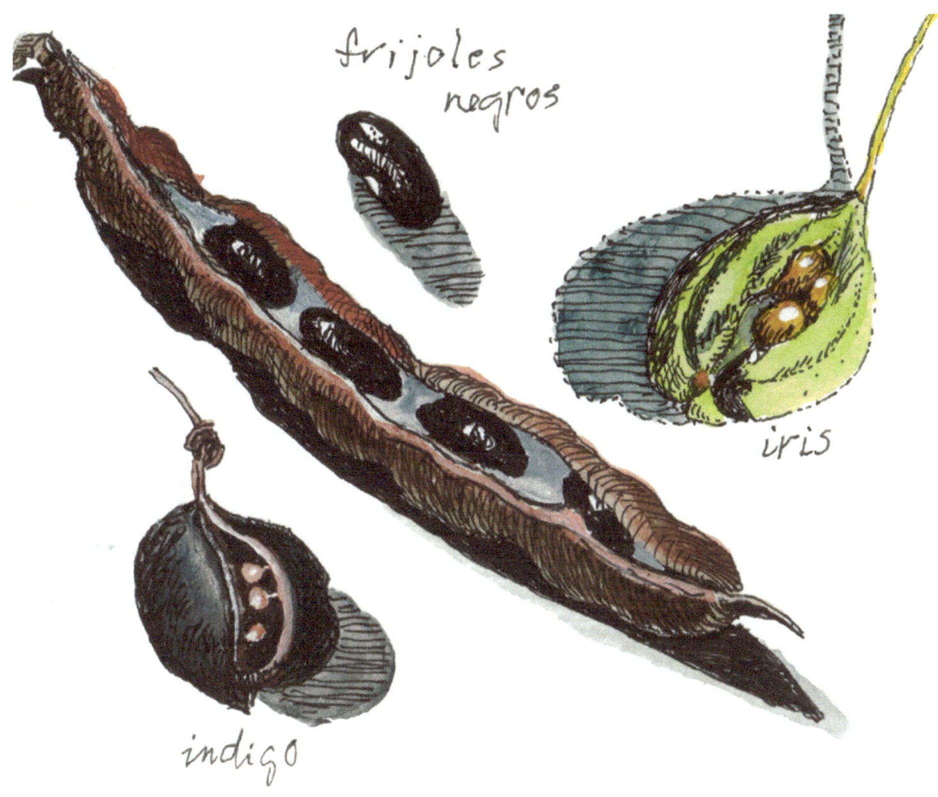

pod has not developed its tannins yet as it is still filled with green chlorophyll. As the late summer advances, I expect this pod will turn brown and dry up, then gradually release its seeds, which will easily split open.

FIGS

Figs (*Moraceae* family) are not technically fruit but are actually a collection of inverted flowers; each of the many flowers, if pollinated inside the fig, produces a single, one-seeded fruit, which accounts for the nice crunchiness of figs. The many tiny flowers are pollinated by very small fig wasps, and all this takes place inside the pear-shaped sack that we call a fig.

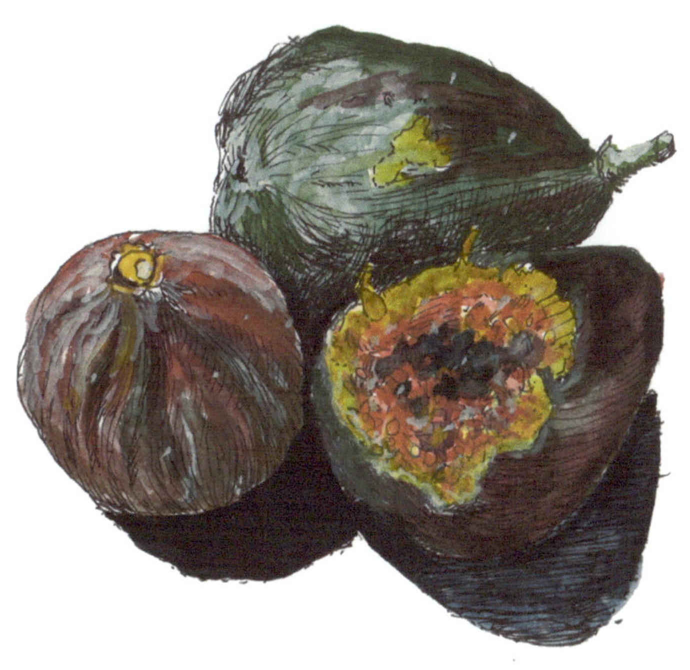

The wisdom of the plants: even when they have roots, there is always an outside where they form a rhizome with something else — with the wind, an animal,

human beings (and there is
also an aspect under which
animals themselves form
rhizomes, as do people, etc.).

Deleuze, Gilles and Felix Guattari, A Thousand Plateaus:
Capitalism and Schizophrenia, *Minneapolis, MN:
The University of Minnesota Press, 1987, p. 11.*

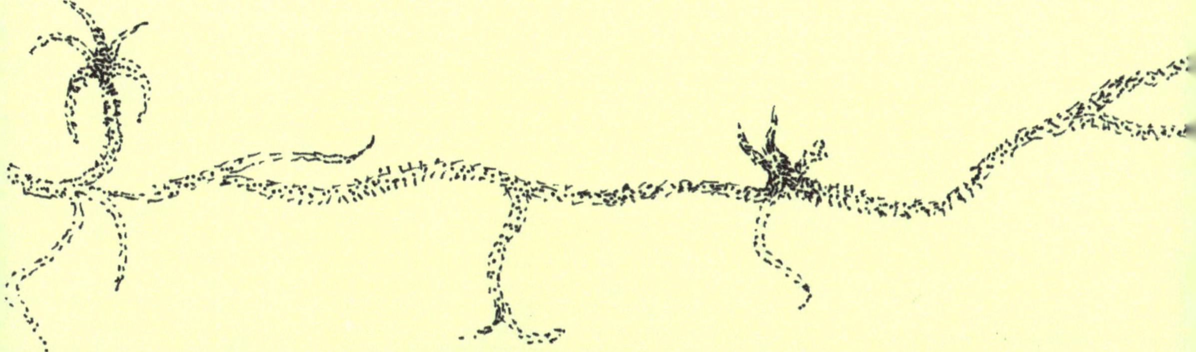

GALLS

Here are various strange little temporary planet-like swellings: on the left are a couple of mature rose hips along with one still in process under its once-pollinator-alluring flower. Inside the hips are the real business of the rose: seeds. As the hip swells, its little defensive thorns fall off, leaving the seed container itself easy to get at for birds and humans attracted to its sweet vitamin C-rich fruit; after ingesting the fruit, of course the birds poop out the seeds as their tough seed coats render them indigestible.

On the right are two stages of an acorn plum gall I found under an oak tree on my road. At first it was reddish with brown patches and firm and heavy like a plum. This gall developed on the scaled cap of an acorn after a tiny cynipid wasp laid an egg in the cap. As the minute egg hatched, it secreted an enzyme that spurred the oak tree to grow an enclosure around the larva.

After I put the gall on my drawing table it began to develop tannin and turn brown as it wrinkled a bit from being cut off from tree moisture. I cut it in half and found a perfectly round chamber in which several minute larvae lay. The acorn plum gall wasp laid only one of the eggs; other female wasps added to the gall apparently; hence, there were several 2mm long larvae inside.

The larva/e feed on the gall until it or they metamorphose into pupae and then new gall wasps. Galls and gall wasps are good food for birds, and gall wasps pollinate figs! When humans came along, they discovered that galls could be crushed and made into good medicine to heal eye infections as well as cuts and burns.

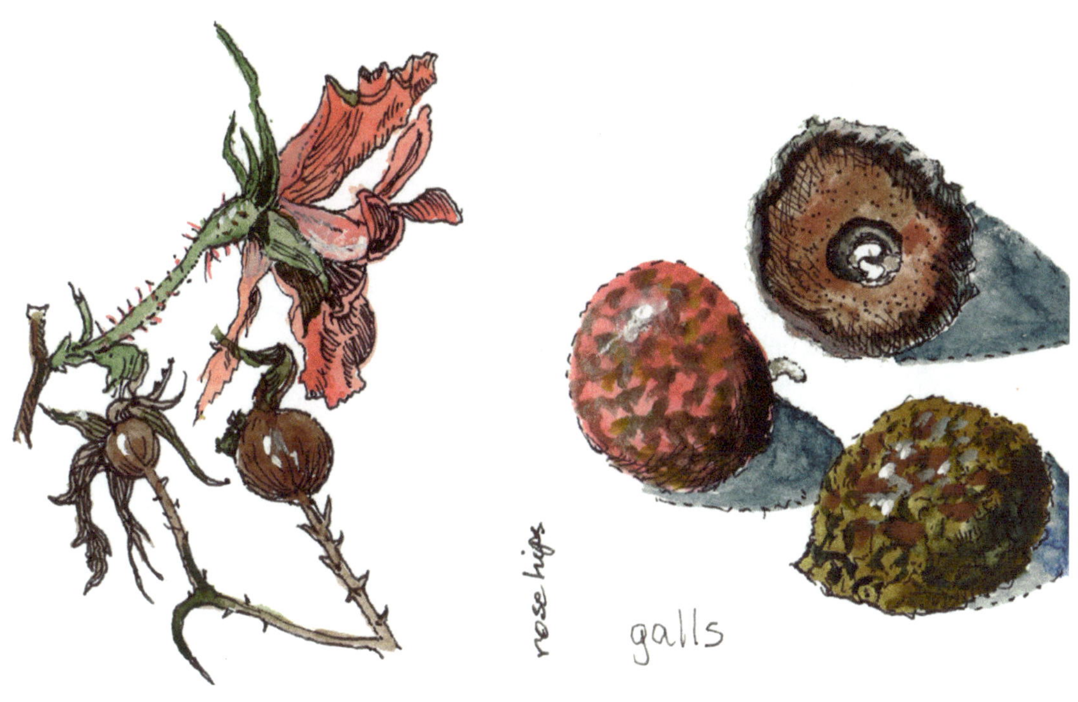

In Medieval times, humans used the tannin-rich galls to make long-lasting ink. You can still make lovely ink by crushing galls and boiling them in vinegar for the length of time it takes to say three Pater Nosters (says the Medieval recipe), then you can add a rusty nail or other ferrous material (such as red earth or a bit of ferric chloride) to turn the ink blue black.

https://www.instructables.com/Making-Iron-Gall-Ink/

YELLOW CHESTNUT TREE -
first buckeye

The other day while hiking the washed-out river trail I looked down and spotted the first buckeye of the season that I've found. As many of you know, I've been obsessively following these things from their pretty yellow and red blooms, through the natural culling of weaker ones as they turned blackish and then vanished, to the plumping out of the yellow ochre and greenish hulls, until the ripening of the stems and dehiscent lines in the hulls. And now the lovely, burnished buck's-eye-like seeds themselves. I collected this one along with the pieces of its split hull.

Nature has built in protection for this seed of the yellow buckeye tree (aesculus flava) by packing it with saponins that are bitter-tasting and poisonous to most animals and most humans. Native people were observant and patient enough to devise a means of leaching out the saponins by boiling the nuts for hours, thereby rendering the nutmeat edible, somewhat like acorns are, as a flour that could be made into meal. They also discovered medicinal uses for buckeye meat. But most humans today treasure this beautiful seed as a good luck charm rather than a food source. And that's lucky for the yellow buckeye tree, whose labors in bringing forth this miraculous thing have been in the service of growing more yellow buckeye trees, and incidentally at the same time providing nutritious nectar for its pollinators.

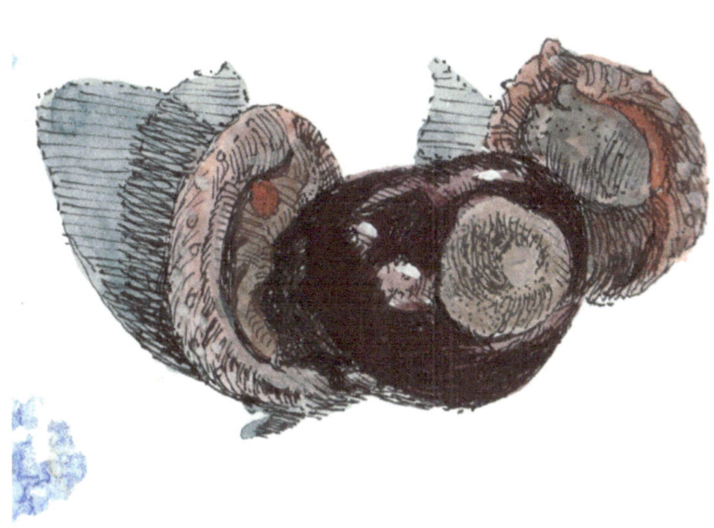

ANEMONE

My friend gave me this luminous anemone with several buds. This is a cultivar that has been named Jules Hubert. Anemones come in numerous colors and varieties, all evolved and cultivated from the simple buttercup (*Ranunculaceae* family). Anemones are poisonous to humans and other mammals but supply nectar and propolis to pollinators, who in turn enable them to make their seeds (which they carry in bouncing lollipop-like balls on the ends of wiry stems, which somewhat look like the flower buds only in a greyish tone).

Humans love anemones for their beauty, and humans inadvertently help them in their project of survival and multiplying by propagating them and spreading them far and wide from their native Japan.

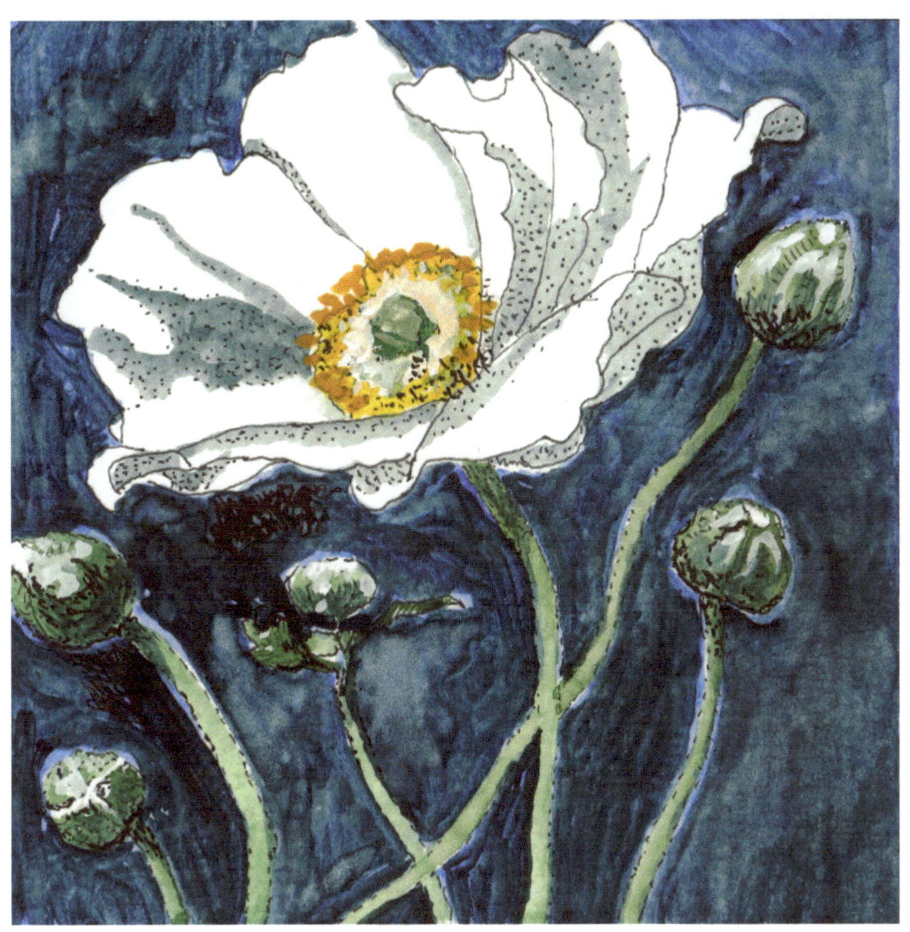

JERUSALEM ARTICHOKES

Jerusalem artichokes (family *Asteraceae* and called sunchokes in the markets here) are blooming like crazy, eight to ten feet tall, working on pollinator attracting and seed making. Meanwhile under the ground their tubers, from which next season's sunflower-like flower stalks will sprout, are storing iron and inulin and potassium and other materials that have made the tubers attractive to humans for centuries.

Native to the eastern part of the Americas, and first cultivated by Indigenous peoples, they do well in Europe also, and in fact, in 2002 the Nice, France, Festival for Heritage of the French Cuisine named Jerusalem artichokes the "best soup vegetable." Roast them with other root vegetables, slice them this for salads, add them to pureed soups after steaming, or ferment them to make brandy as people have done in Germany.

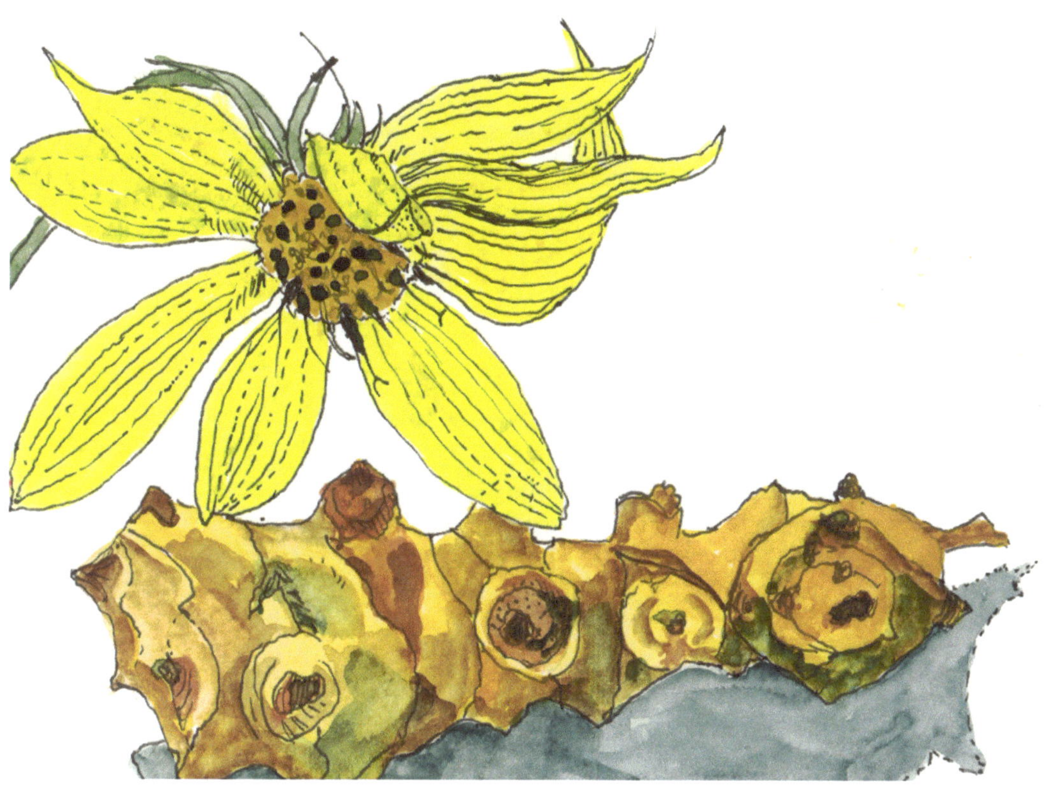

GOLDENROD

I saw a golden glow through dense fog at the park at daybreak the other morning: goldenrod! The plants are growing all around the edge of the woods that surround the ponds. I've heard many people complaining about goldenrod causing allergies, so I decided to get up close and learn about it.

What I learned is that goldenrod, long used as a medicinal for numerous afflictions and ailments, is actually not the plant that causes so many people to sneeze in late summer and fall. Goldenrod produces heavy, large pollen particles that easily attach to the bodies of its many fans/pollinators, such as bees, butterflies, ants, etc. On the other hand, ragweed, which also blooms at this time of year and in the same places, spreads its tiny, numerous, light pollen on the wind, and is therefore the cause of all the sneezing that goldenrod takes the blame for.

Both plants are in the *Asteraceae* family, and both have been used for millennia as medicinals as well as making seeds to maintain their species.

https://en.wikipedia.org/wiki/Goldenrod

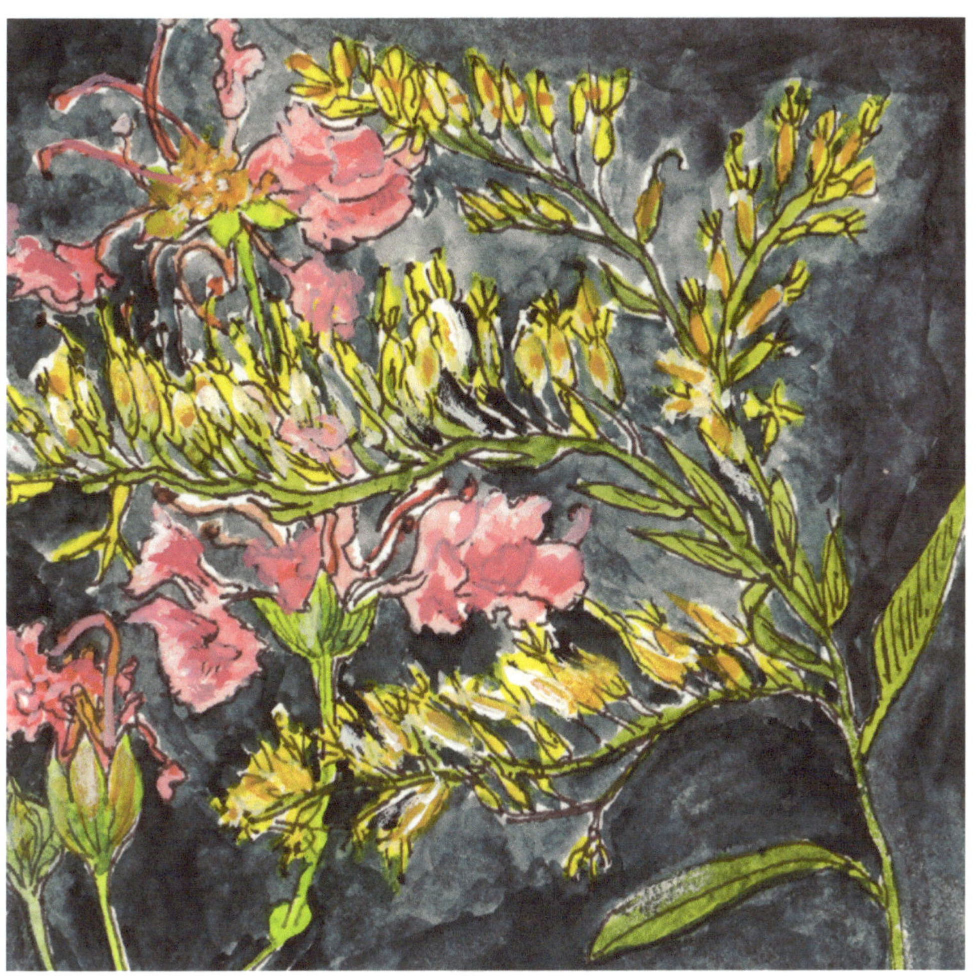

MORNING GLORY -
blasting open

Things are blasting open, moving apart, floating on the frequent updrafts, drying out, and falling in damp layers on the forest floor. Today's gleaning includes a woodland sunflower (*Asteraceae* family) with its seed head bristling with tiny black seeds and its two remaining spent petals ready to join others in the compost below the plant.

The morning glory seed heads are covered with feathery hairs. Once the somewhat hallucinogenic seeds pop out of their capsules they fall to the ground. Native Mexicans used the seeds for their hallucinogenic properties in some religious ceremonies. The Japanese and Chinese who first cultivated the plant used it in their pharmacopeias as a laxative. Morning glory vines and leaves are food for many insects, and pollinators, especially hummingbirds, feast on the flowers' nectar.

An interesting bit of morning glory early history is that early Aztecs added morning glory juice to rubber tree sap to make rubber balls that could bounce, a vulcanizing process that pre-dated Charles Goodyear's vulcanization process by some 6000 years.

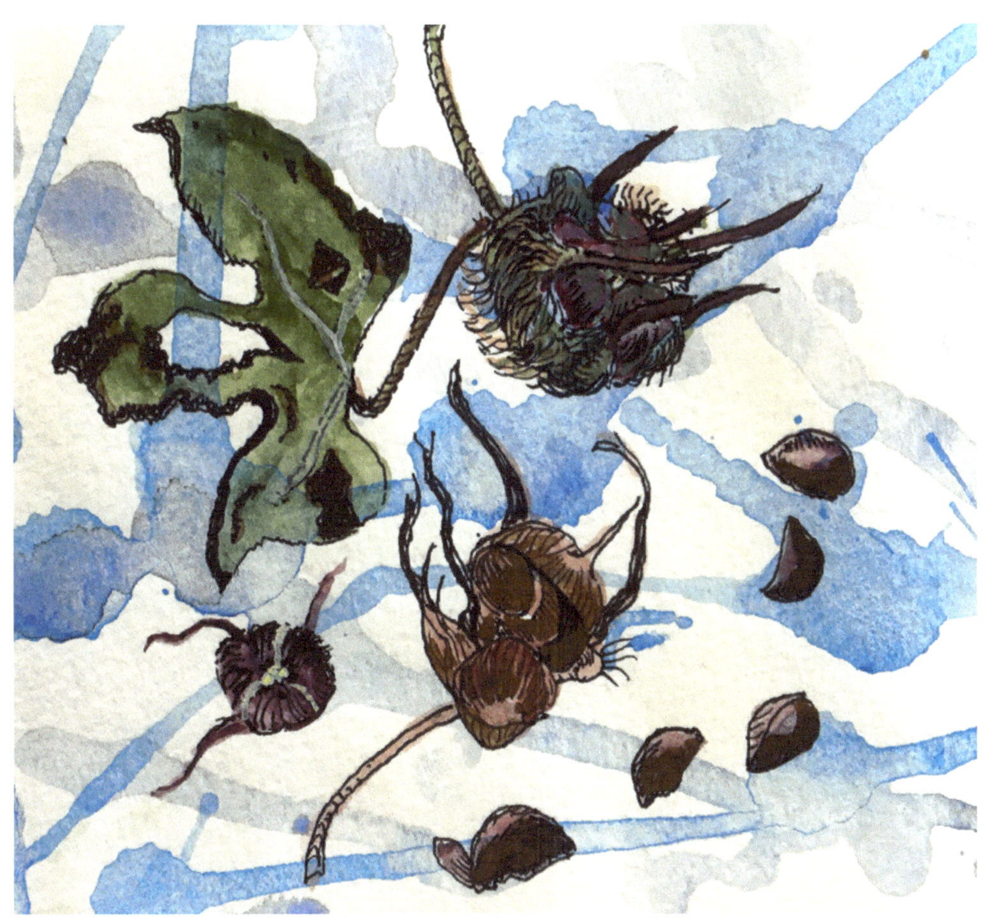

TULIP POPLAR SEEDING

How to let go- Here is a tulip poplar cone releasing her babies, trusting them to find their way. These winged seeds can fly far away or drop to the base of the tree and hang around for up to five years or so before they sprout (or not). Squirrels and birds might eat them, then carry them far away before pooping out the seeds, which can then put down roots in a different time zone.

Tulip poplars are the tallest trees in the eastern forests. The one on our road is 200 feet high by my estimate. Their tulip-like blooms exude a sweet, somewhat watery nectar that butterflies and bees and other pollinators love. If you find a bloom low enough on the tree you can tip the nectar into your mouth. It's also possible to collect the nectar, boil it down a bit to condense it, and make poplar honey or syrup.

My own favorite use for this tree is to use its lumber for woodblocks. Although classified as a hardwood, it has a grain that makes it excellent for carving. I personally use it to make relief print blocks, especially ones that need to stay unwarped for a long time. In 2007-9 I used tulip poplar lumber dried to 5% to make 57 woodblocks for a book project for which I had a grant. The blocks had to remain dimensionally stable for the two years during which I was carving and then printing them. Each page of the 25-page book involved two or three blocks printed in perfect registration, so I needed wood that would not warp or shrink.

Tulip poplar (*Magnoleaceae* family) has been around for a long time. First Nations people have used many parts of it for various kinds of medicine. Many animals forage on the leaves and flowers and cones.

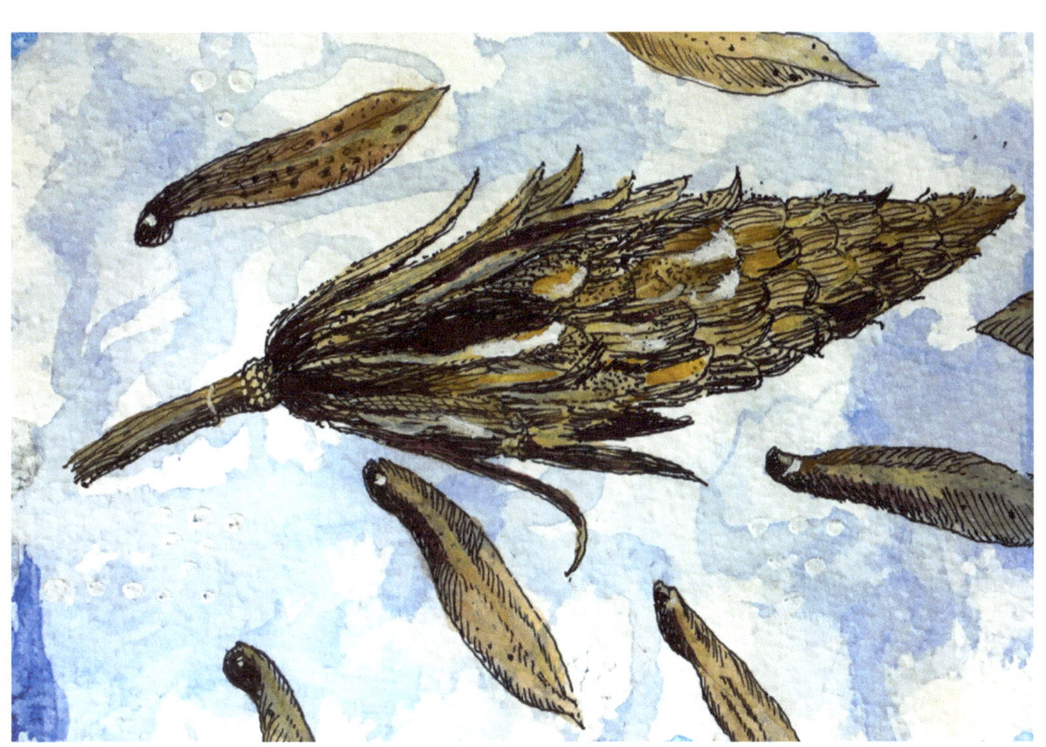

TULIP POPLAR CONES

After all the seeds have left, the mother stands alone. I can't get over
the elegant architecture of this poplar cone. Starting on the left, this
is the cage-like structure that the samaras (wing-like seeds) form
encircling the now-dry pistol or long neck of the flower's ovary.
As the samaras are detached by the wind or by birds, the pistol is
increasingly free until on the right she stands alone. I like how her
once-smooth surface bears scars from her samaras while she stands
firm. If you look high up in a tulip poplar tree on a bright day after
the leaves have fallen, you can maybe see some of these standing on
through the winter.

https://www.uky.edu/hort/propagation-tulip-poplar

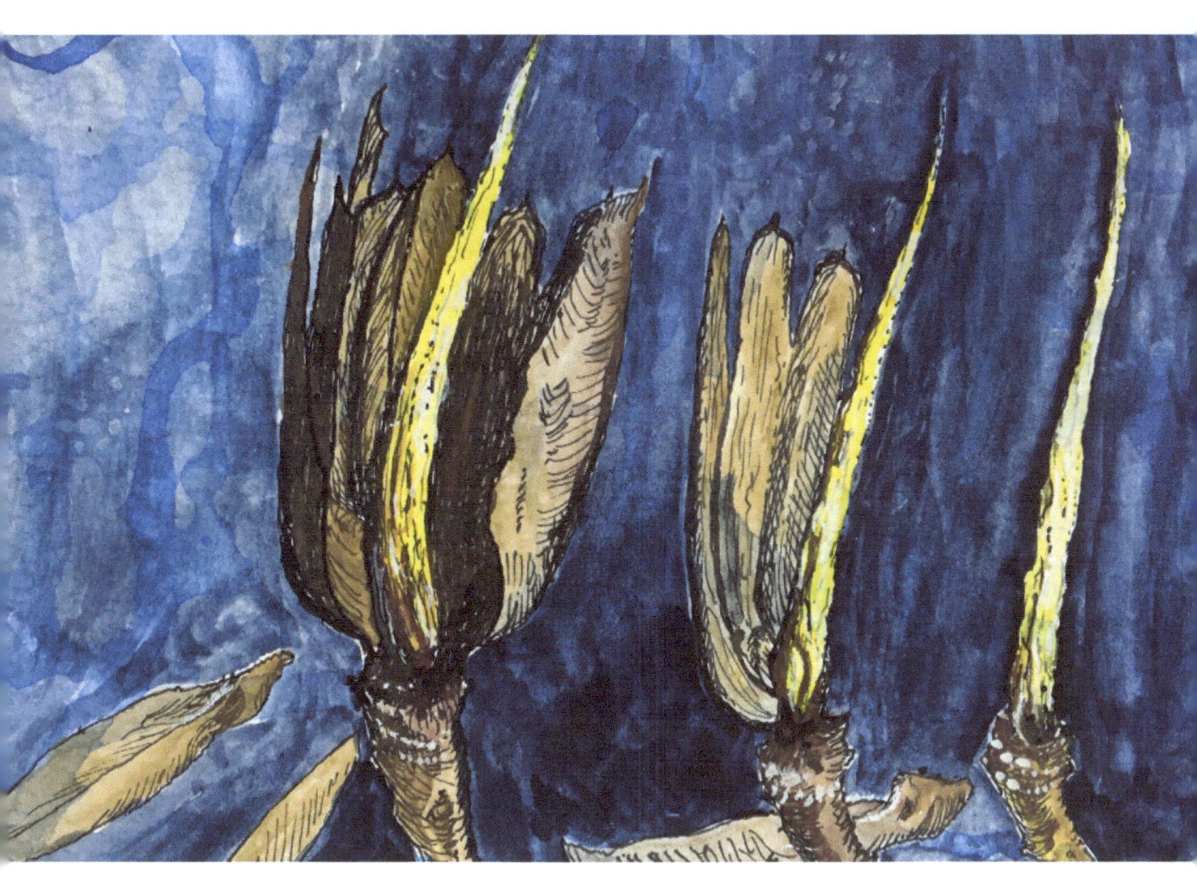

The orchid deterritorializes by forming
an image, a tracing of a wasp; but the
wasp reterritorializes on that image. The
wasp is nevertheless deterritorialized,
becoming a piece in the orchid's

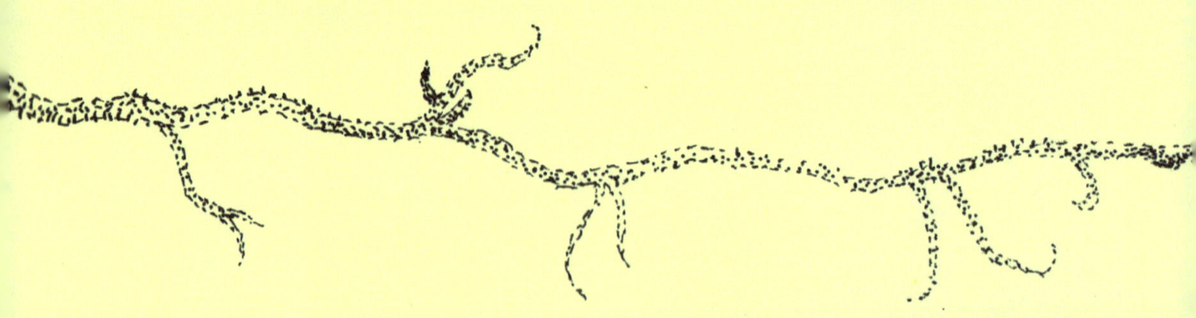

reproductive apparatus. But it reterritorializes the orchid by transporting its pollen. Wasp and orchid, as heterogeneous elements, form a rhizome.

Ibid. pg 10

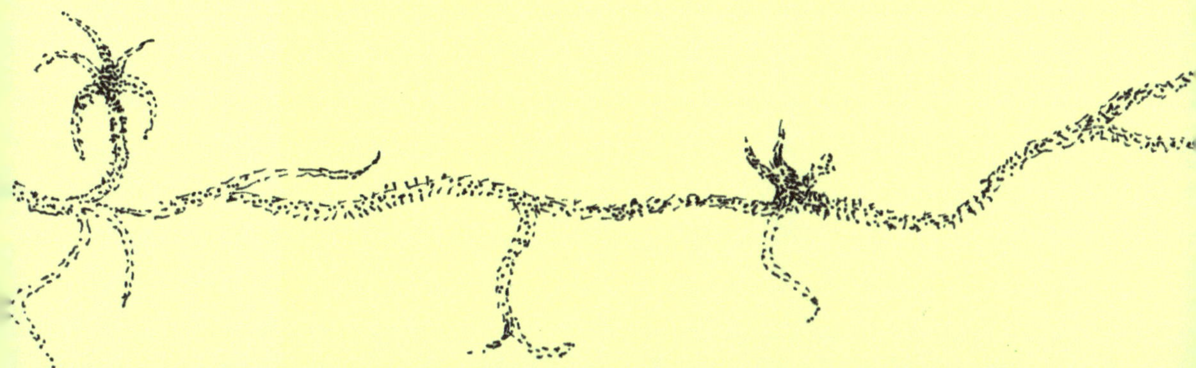

CHESTNUT TREES
– a Chinese-American collaboration

Chestnut trees (*Fagaceae* family) have been something of a mystery to me. When we moved to NC from the Midwest in 1984 we bought an early 20th century house with wormy chestnut built-in bookshelves and interior doors. The realtor was very excited about this wood and explained that until 1904 the American chestnut tree had been the predominant tree in the eastern US, but that it was virtually all gone thanks to a chestnut blight introduced from Asia.

Occasionally I would notice a tree with bristly burrs in late summer, once in a park in New Jersey, another time in New Hampshire. I once bought a bag of chestnut or castagna flour in Italy, mostly out of curiosity, and made a heavy cake out of it. But until a few weeks ago when a friend gave me a bristly Chinese chestnut burr with three nuts inside its opening, I didn't pay much attention to this apparently extinct tree. I started sketching it and wondered how this tree had escaped the infestation of whatever had wiped out the American chestnuts.

Then the other day I hiked up to a high meadow and found two large chestnut trees in the center clearing with dozens of opened burrs under one of the trees. The burrs were bigger than the Chinese chestnut, and the relatively small nuts were lightly covered with fine pale hairs.

Could these possibly be remnants of the American chestnut forest that once dominated the mountain landscape here? My friend

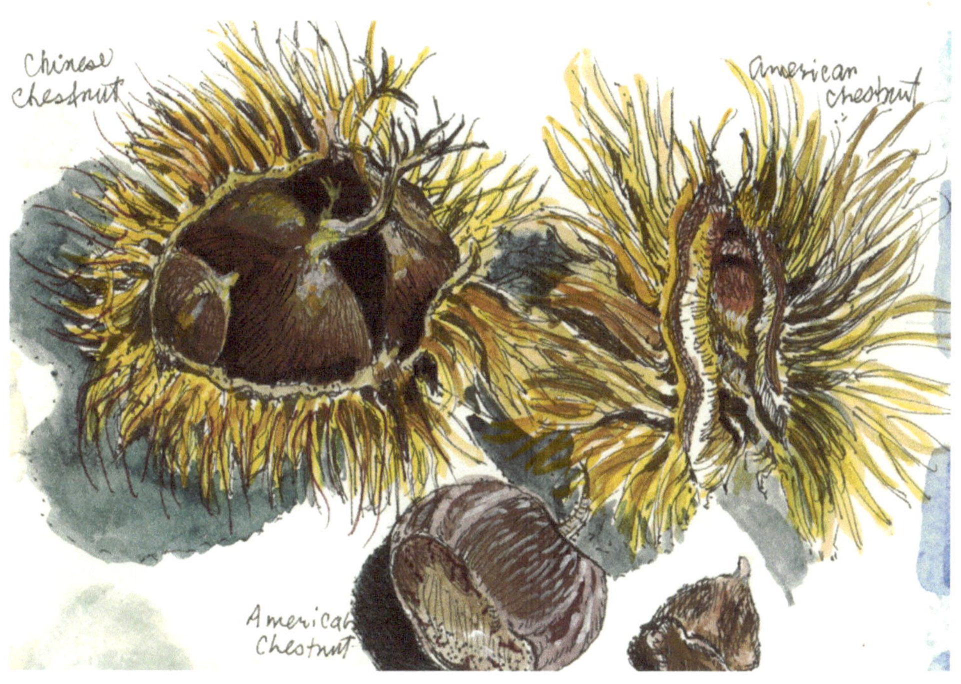

thought they might be Chinese chestnuts, but the softly fuzzy nuts and other features (smooth dark reddish brown stems and twigs along with hairy nuts as well as a few other features) convinced us that these two are American chestnut trees.

Here's a site: https://acf.org/resources/identification/

ECHINACEA AND CHESTNUTS
conferring dignity and power ~

On the left, an echinacea (aka coneflower, family *Asteraceae*) seed pod sits in all its life-containing elegance on the head of a small discarded Victorian china doll. On the right a different china doll, also a factory discard, is elevated by wearing an American chestnut burr, its silky fur lining gently embracing the head of the doll.

Echinacea, as you may know, is native to the plains of North America. For thousands of years numerous groups of Native people chewed and sucked on the root as well as made decoctions and salves and tinctures of this plant to cure maladies ranging from insect bites to rheumatism to snakebites to all kinds of infections. It fell in and out with the reigning medical industries once the invading Europeans introduced it to Europe in the 17th century.

Like all plants, it seeks to reproduce, and it meanwhile provides food for gold finches and other birds and pollen for butterflies and other insects. Unfortunately, today it is classified as At Risk due to human interference by way of habitat destruction, over-harvesting, and pesticides.

For much more information, see https://en.wikipedia.org/wiki/Echinacea_laevigata

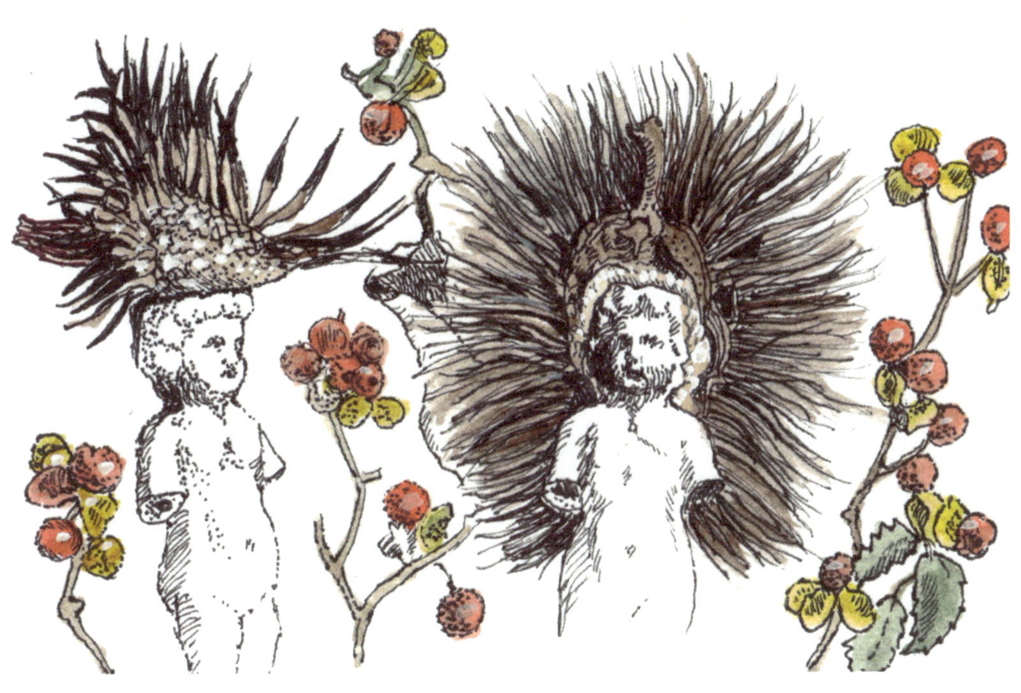

DIAGNOSTIC MEDICAL DOLLS ELEVATED

A few years ago one of my sons and I dropped into an estate sale down the road from his house in rural New England. My eyes went straight to a couple of tiny figures casually lying on top of an old book. "What on earth—?" I asked the seller. "Victorian doctor dolls," she said. She went on to explain that when women had problems with their bodies, the doctor would produce a little androgynous doll from his kit and have the woman point with a tiny stick to the problem area on the doll's body. That way no embarrassing encounters needed to take place.

These lucky medical dolls have headdresses of leather leafed mahonia bloom buds. (Mahonia bealei). Early this morning, while our front yard was under a silvery frost blanket, I spied the mahonia bush busily putting out flower buds. This bush with its evergreen spiked leaves thrives in the low light of winter and puts out its yellow flowers in late fall/early winter. It's practically the only food source for bees around here in winter, and they buzz and bumble around the blooms as snow falls all around them.

The bell-shaped blooms are pollinated by the bees, and by late winter they morph into berries, good food for any birds who are hanging around, too. And humans for thousands of years have eaten not only the berries in pies and jellies and fermented into wines, but also the flowers as salad when there are few other fresh plant parts around. The roots have been used medicinally by Native people to treat tuberculosis, fevers, coughs, arthritis, weak knees, and dysentery among other ailments.

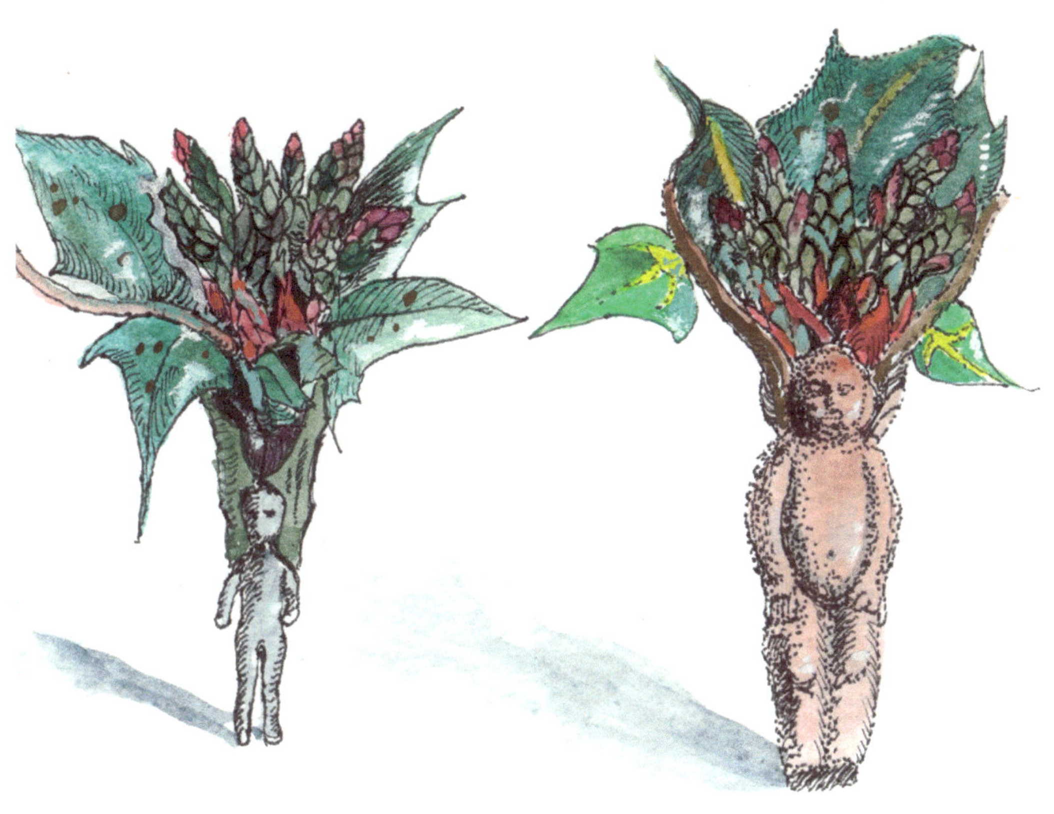

I like to imagine these tiny doctor dolls carrying the cures for the diseases on their heads.

ROOTS

Here are some of the ways that roots do much more than plug plants into the ground: many of them form commensal relationships with the roots of plants of different species; for example, the root-like mycorrhiza of russela mushrooms connect with the roots of pine trees in order to exchange nutrients made by the chlorophyll-filled pine tree for enzymes that the tree needs to better absorb nutrients from the soil. Meanwhile Indian pipes, which also lack chlorophyll, exchange enzymes with the mushrooms to get food in exchange for giving the fungi increased ability to break down lignin and cellulose and the ability to exchange nutrients with the pine tree.

Root hair cells secrete acids that release minerals from soil as the root hairs push their way through the soil. Root tip cells can also nourish bacteria that help ward off rival plants and insects. Roots have been known to help weak neighbors by sending enzymes and nutrients to help them ward off invasions.

Humans know well that some roots, such as those of turnips, store food, mostly in the form of starches. Other roots, horizontal rhizomes, such as the giant bamboo rhizome pictured here, spread the plant by sprouting roots and buds from nodes as they race along slightly below the surface of the soil.

Some roots grow happily with no soil and absorb nutrients from water and even air. There is even evidence that plants communicate and share more than nutrients through the massive entanglement that throbs and hums along silently (to us) beneath our feet.

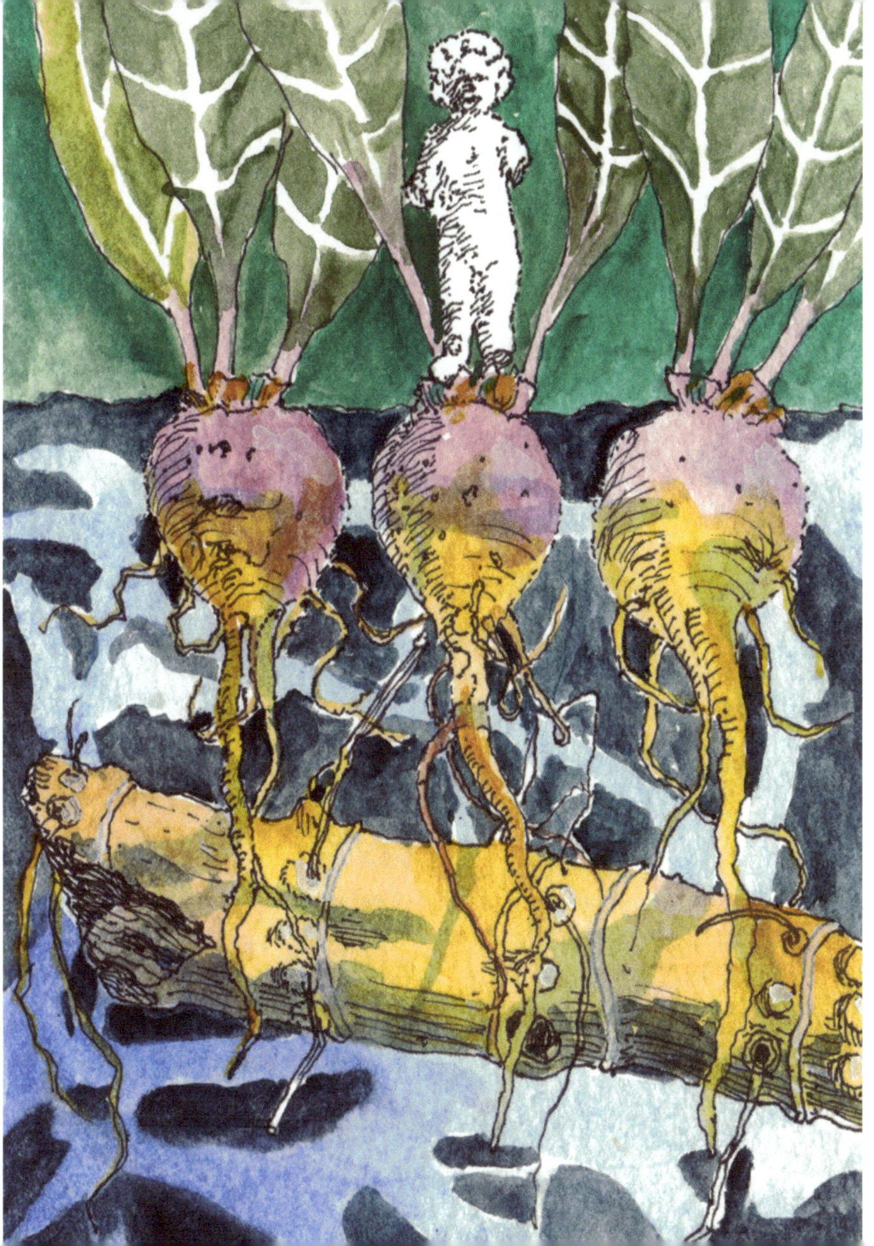

ORCHIDS

A good friend and I went to the greenhouse that was part of a major tourist attraction estate here. It was all poinsettia-ed up for the holidays. Exotic tropical orchids, in strenuous, chemical-fertilizer-induced bloom, were interspersed in their own plastic pots glowing with vivid color.

The *Orchidaceae* family seems to have been around for some 450,000,000 years. Orchids need pollinators in order to naturally re-produce, and the pollinators get nourishment from the nectar they get from the flowers. Each kind of orchid lures its very specific pollinator by either having a form or a scent that fools the pollinator into trying to mate with it or by luring pollinators to it and then trapping them so that they pick up pollen and are then released. Specific orchids are affected by species changes in pollinators and will evolve to become alluring to the changed pollinator. And if humans eradicate a pollinator that they consider undesirable, orchids that depend on that pollinator species can no longer make seeds.

Orchids form mycorrhizal relationships with the rhizoids and roots of fungi and trees, in which the fungi (such as russela mushrooms) siphon off carbon-rich sugar from the trees (who produce it through photosynthesis) and pass on to the trees phosphorus and nitrogen that they have gotten from soil using enzymes the trees don't have. Meanwhile orchid seeds, which have no energy reserve that they need to sprout, receive carbon from the fungus through underground networks.

Some scientists are beginning to consider the forest to be one giant organism rather than a collection of discreet and independent plants.

https://www.earthdate.org/pando-a-forest-of-one

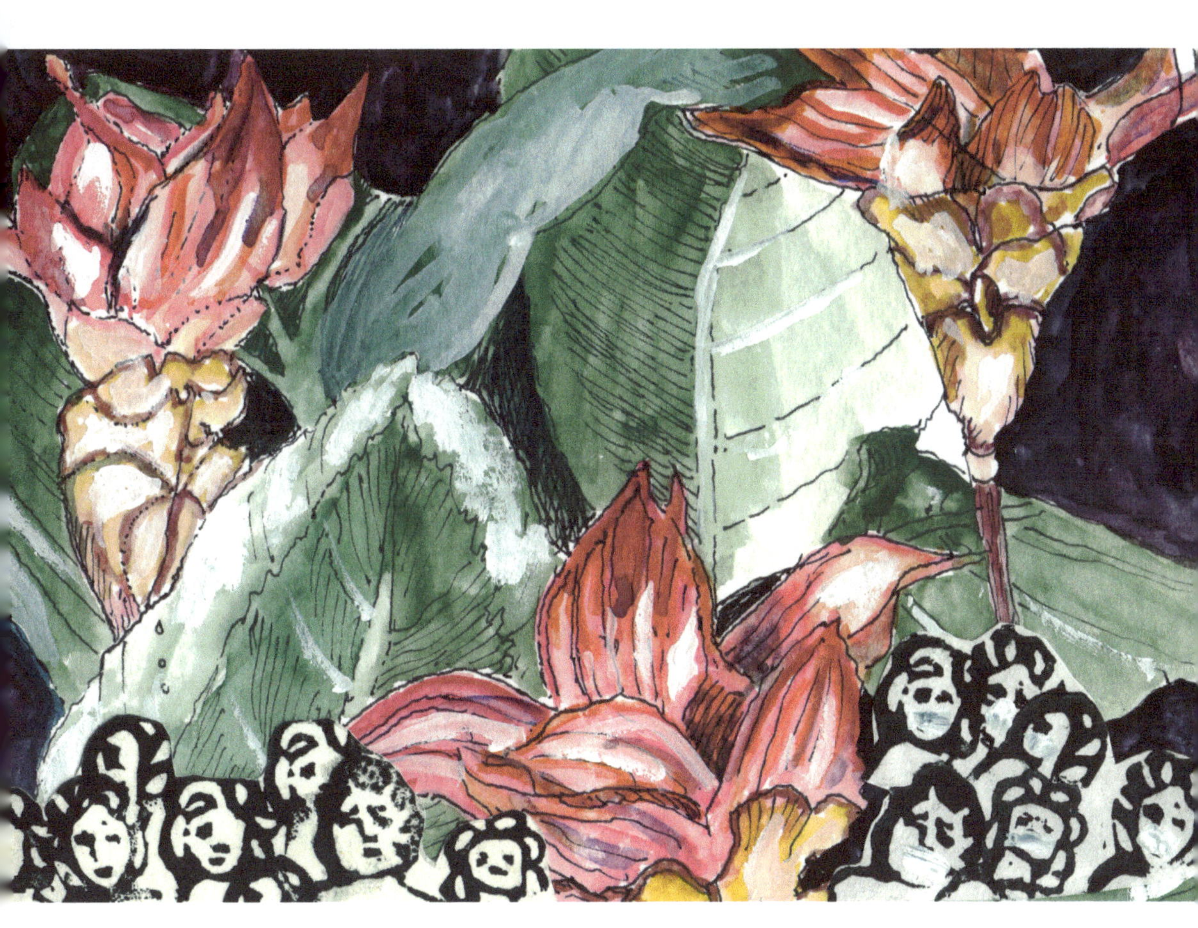

APLECTUM HYEMALE
(from Greek meaning 'spurless' 'winter')

Walking through the forest in late fall or winter you may be lucky enough to see this strange green and white pin-striped single pleated leaf. I found some yesterday in a cluster of seven single-leafed plants, which is generally a sign that this plant, which we commonly call putty root or Adam and Eve, is happily growing in rich soil and in an auspicious setting surrounded by its companion pine trees and russela mushrooms (invisible now but very active underground making soil nutrients available to the plants and trees from which the mushrooms are getting food.) Putty root can make food even now in late fall because it can photosynthesize at temperatures as low as 35 degrees F. The pleats enable the leaves to stay warm-ish by folding themselves close together like old-fashioned plastic rain hats. The strips of chlorophyll on the leaves follow the mountain peaks of the folds so they're exposed to sunlight even as the leaf pleats itself in response to the cold.

This leaf's only job in winter is to make food, which it stores in its underground corm. Botanists refer to the leaf as the male part of the plant (because it has a job?), and a common name for the plant in some places is 'Adam and Eve'. Adam is also busy growing a second corm (dotted lines) from which Eve will sprout in spring after Adam has decomposed and added his spent leaf to the other decomposing leaves. Eve (designated the female part of the plant, although it bears both male stamens and female pistols) will become a tiny orchid whose flowers are pollinated by bees and are also self-pollinating. Its tiny seeds are dispersed by the wind.

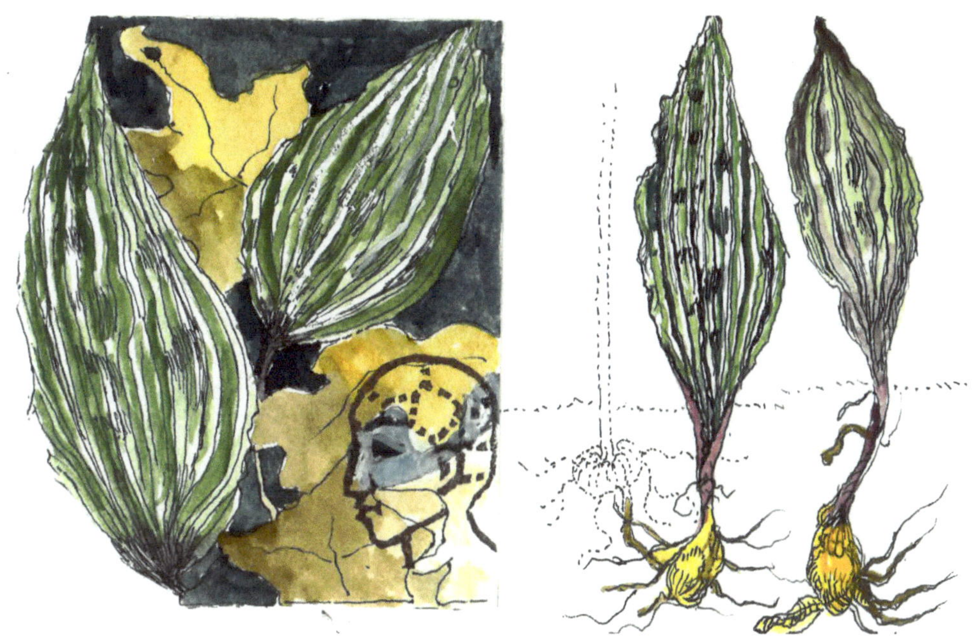

The name putty root came about because Native people discovered
that the mucilaginous juice of the corm was good for mending pottery,
information that they passed on to European colonists. My friend and I
dug a corm a few years ago and macerated part of it and used it to mend
a small piece of pottery. The juice also made an excellent water-soluble
paper adhesive.

AGAVE

Agave (fam. *Asparagaceae*) is seeding across the street in a neighbor's front yard here in the humid Appalachians. Such a displaced plant, originally from southern Mexico and beloved by the Aztecs and Mayans for thousands of years for its fibrous leaves, with which they made rope and twine, rough fabric, paper, and even medicines and food.

The many varieties of agave migrated and naturalized to the southwestern desert areas of what is now part of the USA; and then the plant was appropriated for capitalist ventures when humans found it to be an attractive landscape plant as well as the source of juice that could be fermented and made into mezcal and tequila, and the fiber – which is now called sisal – into binding twine, rope, carpets, paper, and even dartboards.

Agave is pollinated at night by long-nosed bats, a rare kind of bat that has evolved to have a long nose and tongue that can reach into the tubular flowers for nectar, while picking up pollen on its fur. There are also a few bees that seek out its nectar. When humans grow factory farms for making tequila from the agave nectar, they take over pollination from the bats that live in the area, causing starvation of the bats, so that now these bats (leptonycteris) are declining.

Agave plants are sometimes called century plants because it can take up to 60-80 years for an agave plant to store up the energy to send up a bloom stalk. Once the plant blooms, it sets seeds as well as bulbils, and the plant dies. When it's growing in a good environment, young plants then sprout from the roots, bulbils, and seeds.

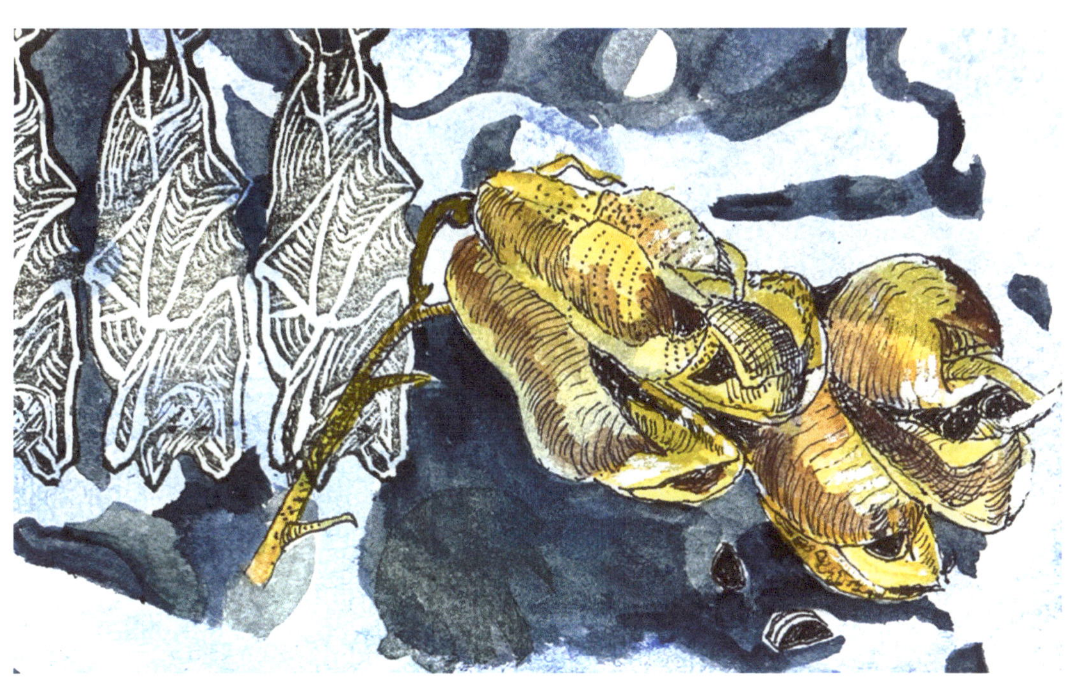

JAPANESE MAGNOLIA BUDS

The Japanese magnolia (*Magnoliaceae* family) is full of fat, fuzzy bloom buds, even though we're still at the midpoint of winter by the old calendars. There is not a single leaf on the tree, but the buds that are starting to open are showing some of the red pigment (anthocyanin) on the inside of the tepals. This pigment helps a plant cope with cold temperatures. The stems near the buds are also that deep reddish maroon.

Magnolias are among the oldest angiosperms (flowering plants), appearing in 100,000,000 year- old fossils. Even older by another 150,000,000 years are the beetles that evolved before and along with them and are their main pollinators.

There are several species of beetles that seek out the nutritious pollen of the magnolia (and that evolved long before butterflies and bees) as their most important food. But beetles are clumsy pollinators with mandibular jaws with which they chomp through the leathery tepals in order to find the coarse pollen. The seed-producing stamens and pistols are fortuitously sturdy enough to withstand the crunching and bumbling of their beetle partners. The resulting seeds are protected in cone-like pods.

The Japanese magnolia blooms in winter around here when freezing weather is a regular occurrence, but the beetles are equipped to come out of their winter diapause: they have evolved ways of coping with temperatures way below freezing, including getting rid of all the water and food inside of their bodies so that their cells won't freeze. They can also produce antifreeze chemicals such as glycerol and sugar.

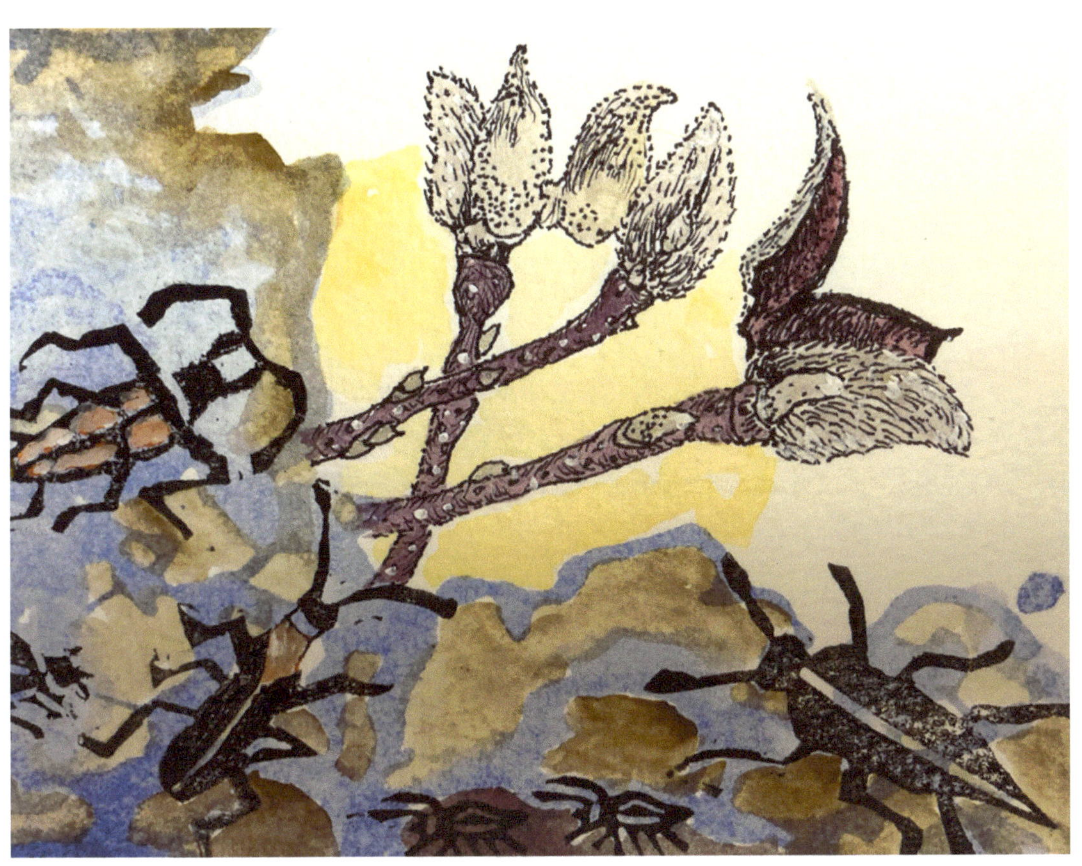

So tread carefully if you have a Japanese magnolia tree under your care.
One of the most-asked questions I found while researching magnolias was
"How can I get rid of the beetles that are on my magnolia tree?"

DECEMBER MOON
and Leather Leafed Mahonia

The leather leafed mahonia (fam. *Berberidaceae*) bloom buds are developing beautiful crimson patches at the base of each bud. It's surprising to see this fresh glowing color appearing and developing when everything else around the woods and gardens is drying up, showing the decaying browns of tannin, and falling off. But mahonia is getting ready to burst into bloom, and to do this she needs the help of the red pigment anthocyanin. This pigment helps the plants withstand cold as well as screens out damaging ultraviolet rays and air pollution, thus giving the plants the extra boosts they need to produce seeds in winter.

The waxing crescent moon is still the bear moon of December, and as it changes its position relative to our section of earth it affects plants as well as fish: just as the moon's gravity exerts a magnetic pull on the earth's oceans, causing the tides, it also affects the moisture in soil. Seeds absorb more water during the full as well as new moon because more water comes to the surface of soil. Traditional practices and permaculture tell us to plant above- ground crops during a waxing moon phase and root crops during the waning phase of the moon because of the moon's effects on soil moisture. Fish tend to surface more during a full moon, and some fish rely on the extreme tides at the full and new moons to go to their spawning grounds. Animals such as bears and humans find their best fishing to be at the full and new moons.

To find out more about the Native American moons, go to https:// aacimotaatiiyankwi.org/ecology/lunar-months/mahkwa-kiilhswa/

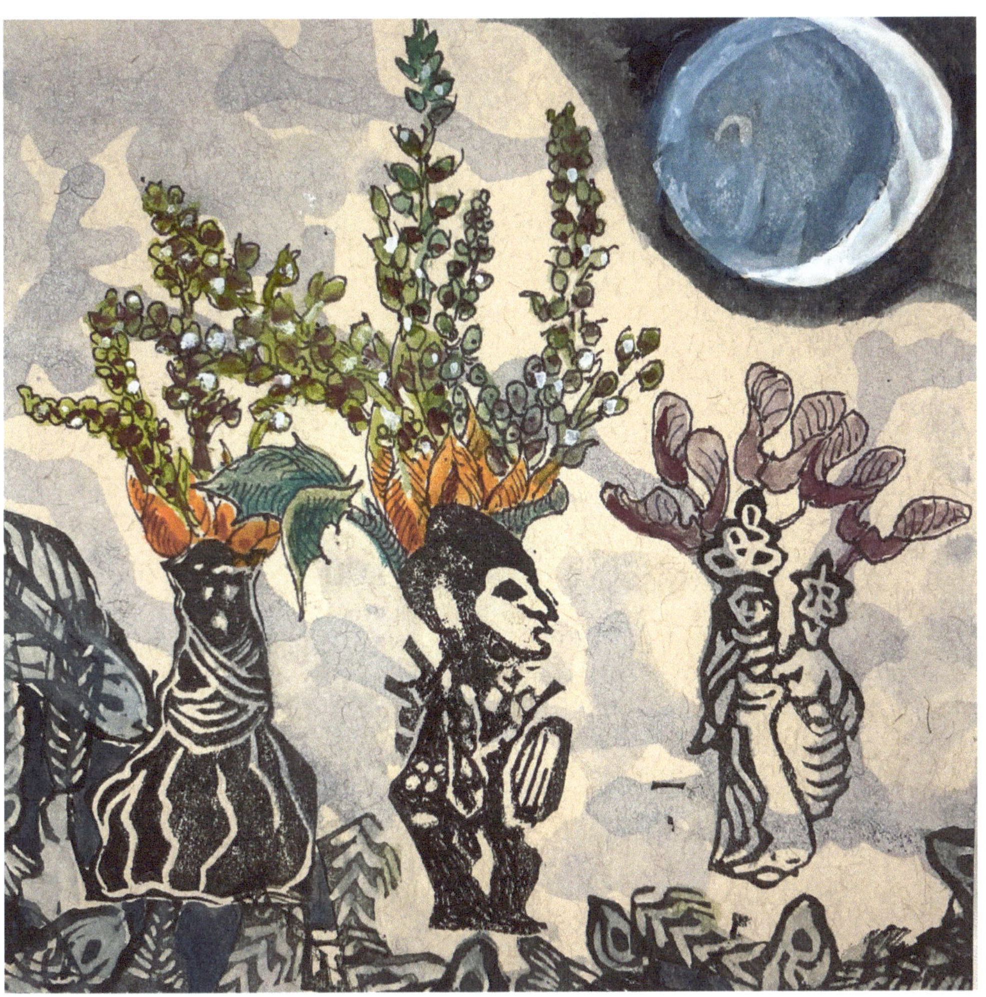

CAROLINA ALLSPICE

And the Solstice has arrived! To celebrate I am presenting a close relative of the ancient magnolias: the also-ancient Carolina Allspice (family *Calycanthaceae*), which involves cantharophily, or pollination by beetles. Like the magnolias, the flower of allspice has tough, leathery petal-like tepals. The whole plant has an enticing spicy fragrance that comes from oils rather than nectar. The smell is irresistible to beetles, who burrow their way into partly opened flowers and get trapped inside long enough to deposit pollen from a different flower. The beetle meanwhile and inadvertently, covers its body with more pollen. At that point the flower releases the beetle to crawl away and spread the pollen load to another flower, and the first flower turns off its aroma and gets busy with seed production.

Nature usually does fine as long as humans stay out of her way. Native people lived respectfully among plants, using this plant's bark as a spice, while recognizing that the seeds and flowers and leaves were protecting themselves by being poisonous to humans and exercising caution.

Carolina allspice makes use of mycorrhizal fungal networks that help it send and receive chemicals to and from chlorophyll-bearing plants as well as help it send and receive chemical signals to and from other plants warning of dangers such as imminent attacks and droughts.

https://www.fs.fed.us/wildflowers/pollinators/animals/beetles.shtml

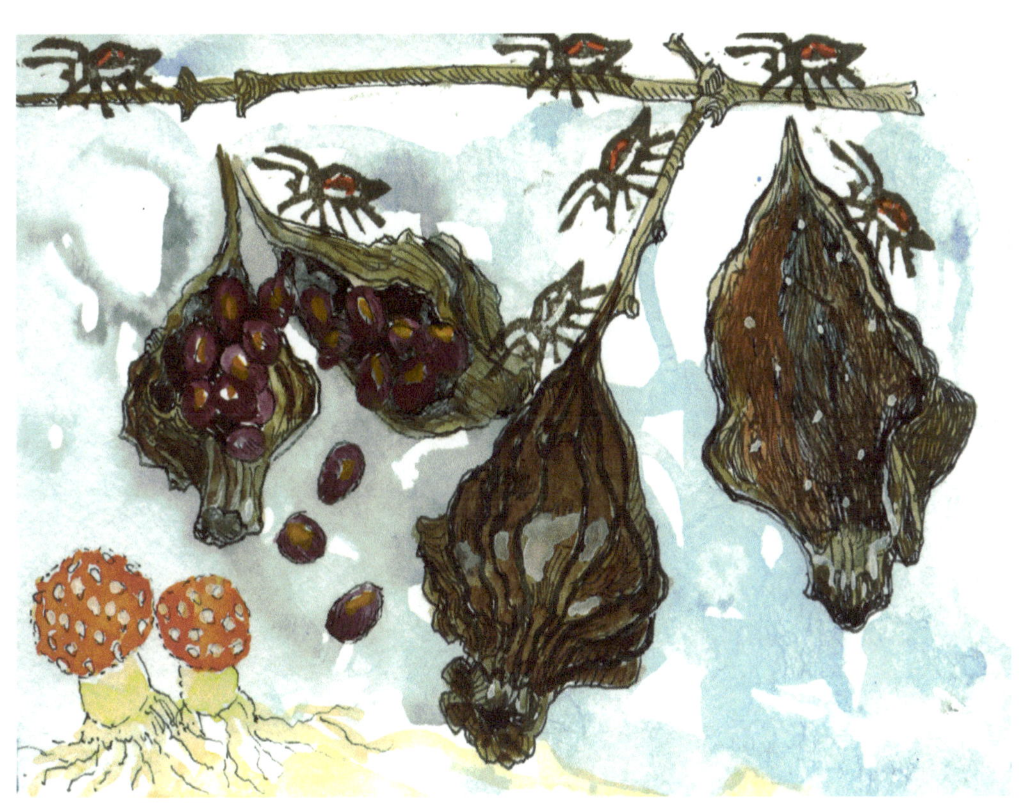

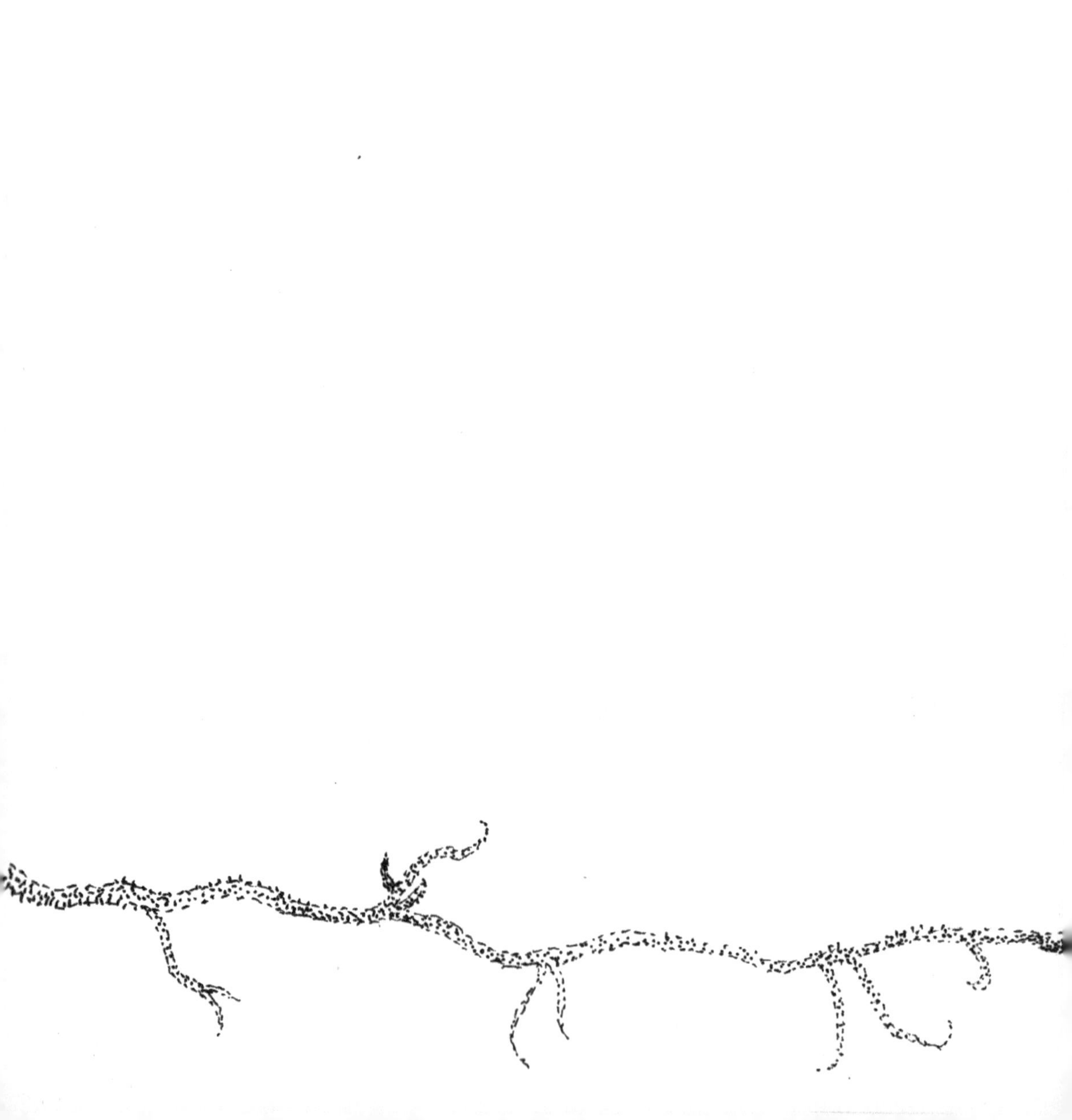

Thought lags behind nature

Ibid. pg 5

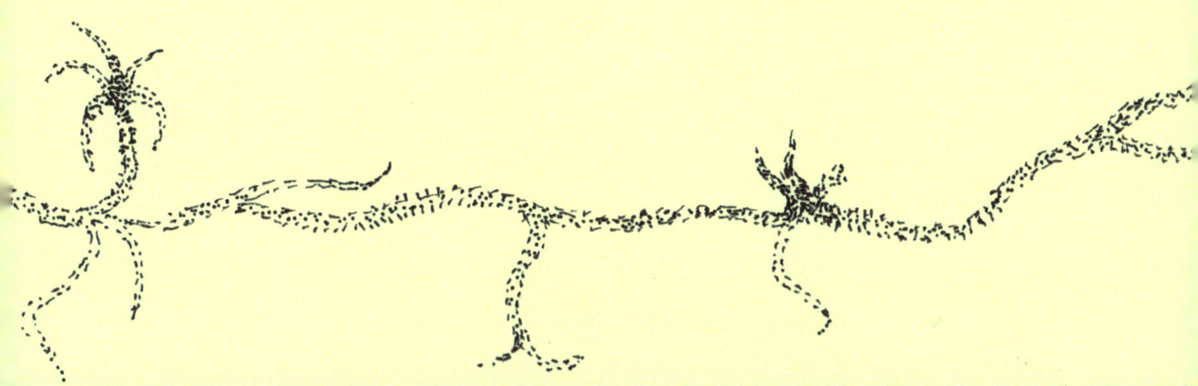

EVENING PRIMROSE

Evening primrose (fam. *Onagroceae*) is seeding in late December around here. I found a patch of them behind an abandoned labyrinth on the edge of the soccer field on campus. I like to imagine the hawk moths and tiny twilight-flying sweat bees drawn to the sweet smells along with the yellow of the blooms in the long evenings of summer.

The evening primrose also interacts with parasitic moths whose caterpillars feast on its flowers (which stay open for one night only) and numerous seeds. Goldfinches are also fond of its seeds, and in the early evenings when the flowers open one by one, some species of butterflies drink their nectar.

At this time of year, the seeds are being dispersed by the wind, and the dried stalks are providing nesting materials and humus to the often bare earth from which the plants are adept at growing. Meanwhile humans prize the evening primrose seed oils for their medicinal uses, most of which were discovered by Indigenous people thousands of years ago.

https://www.mountsinai.org/health-library/herb/evening-primrose-oil-epo

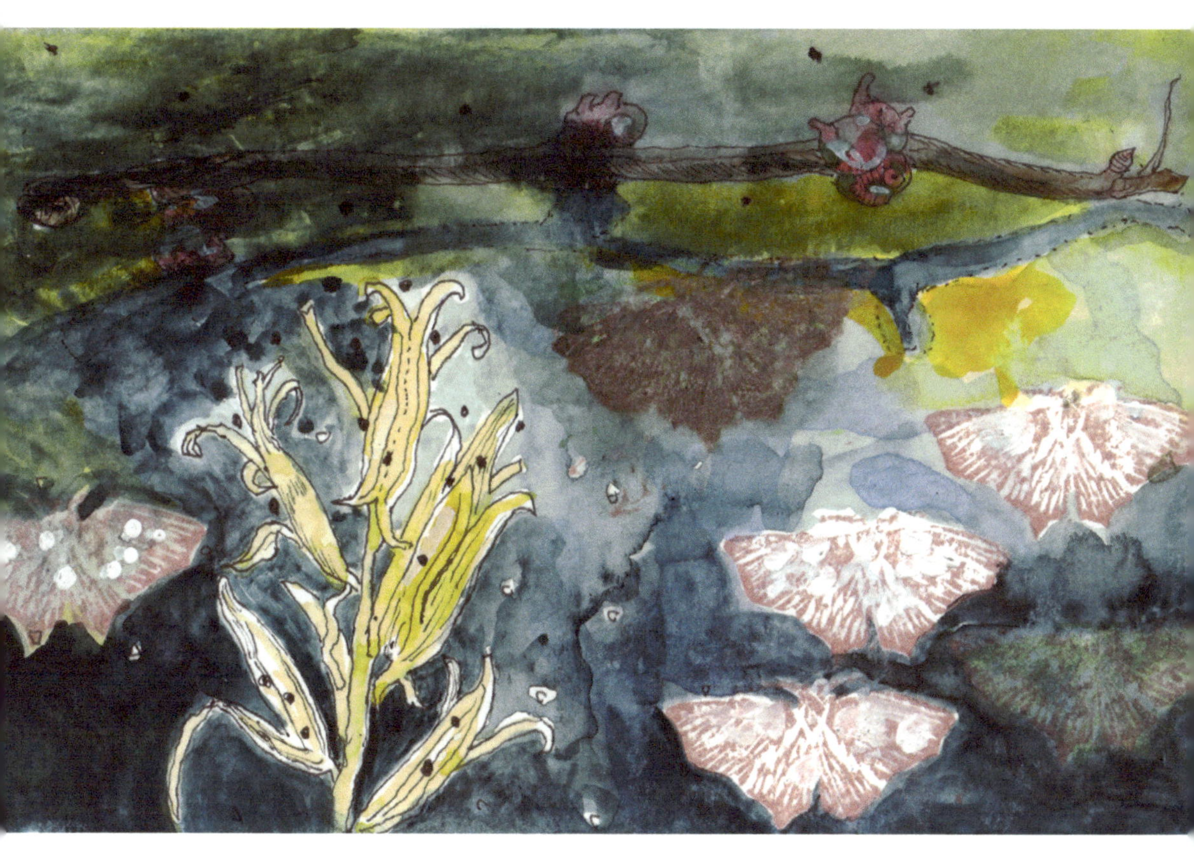

MOSSES

The wind was icy in the woods late this afternoon and there was very little bright color apart from a few leftover bittersweet berries and red rose hips. But if you relax your eyes and crouch down a whole world of bright green comes into focus: it's bryophyte season, with mosses starring right now in our woods. Bryophytes are ancient, and they evolved from algae. They are those strange little fuzzy tree trunk dwellers, that look more like model railroad landscapes than actual living plants. Bryophytes are non-vascular, lacking xylem vessels for transporting water; instead, these mosses and liverworts absorb moisture, carbon dioxide and sunlight directly through their leaves, allowing them to make their own food by photosynthesis.

Some mosses provide nitrogen for the ecosystem and release it into the soil, thanks to their relationship with cyanobacteria that colonize the moss and receive shelter in return for fixed nitrogen and other nutrients. Some mosses anchor themselves to trees and get carbon dioxide directly from the air as epiphytes; but they do not take anything from the trees.

Mosses can live everywhere except in salt water, and they are always ecosystem stabilizes, bringing soil amendments to depleted soils, preventing erosion, and improving soil moisture. They do not have roots; rather, they have rhizoids that anchor lightly to the substrate but that don't suck moisture or nutrients from the soil or tree trunk.

Mosses reproduce by spores rather than by seeds and they are also able to reproduce from cuttings. These essential ancient plants provide insects and small animals with nests and burrows and food. Indigenous people used them for diapers and pillow and mattress stuffing.

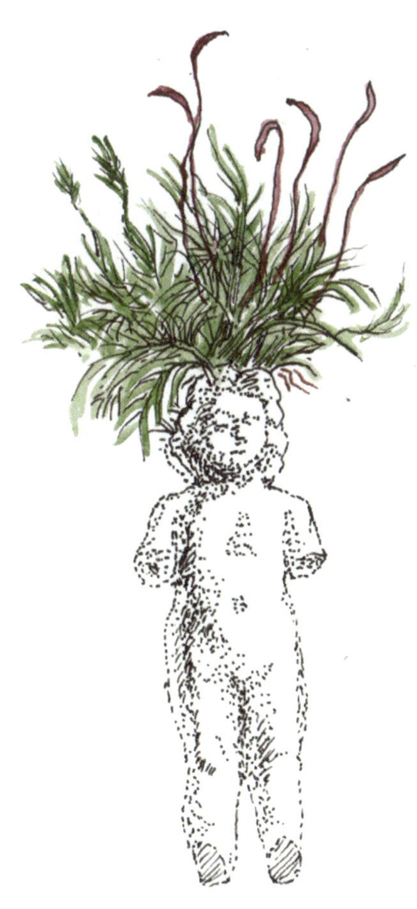

LICHEN

You need to crouch or squat down or press up close to a lichen-covered branch in order to appreciate these wondrous ancient beings. If you're lucky enough to find any, know that the air near them is clean because lichens won't grow in a polluted environment.

A lichen is a synthesis of an alga or a cyanobacterium and a fungus. The algal threads contain chlorophyll and can make food through photosynthesis. The fungal parts absorb and retain moisture.

The two lichens shown here are a beard-like usnea Florida (family *Parmeliaceae*) and a foliose lichen from the *Lobariaceae* family. The usnea has lovely little flat, spore-producing disks and oddly elastic stems. The hair-like structure of usnea plus its strong antibiotic properties and its ability to absorb moisture made it useful to humans as a bandage for wounds as recently as the early 20th century. Lichens have remarkable abilities for surviving in the Arctic, in deserts, and in temperate cities and forests – just about anywhere except in salt water. And in all these environments, they act as filters of environmental pollutants, and they release oxygen into the air. They are slow-growing and can live up to 60 years or so.

One of their coolest tactics is to manage to go dormant when the weather is dry. The outer layer then becomes opaque and grayish. When moisture is again present in the air, the outer layer becomes transparent, allowing the chlorophyll-filled inner strands of algae to become visible and green. The algae strands in lichens need sunlight, and the lichens often grow high up in canopies of trees. Look for lichen on tree trunks and branches as well as on the forest floor after wind.

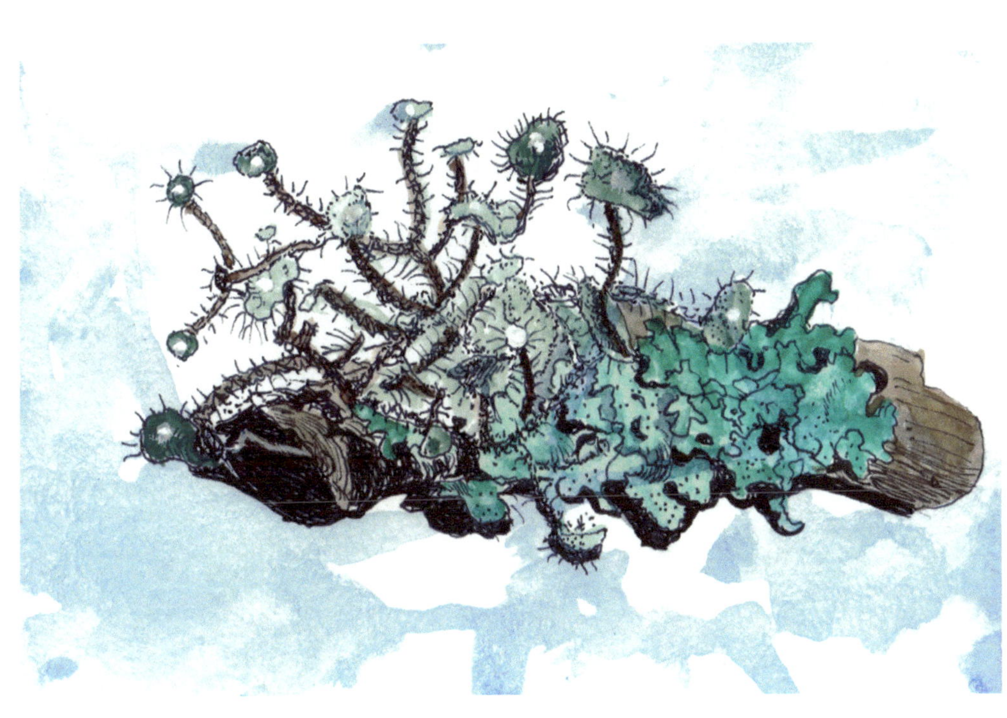

LEATHER LEAFED MAHONIA BLOOM

I drew a different leather leafed mahonia bloom (fam. *Berbereceae*) a few weeks ago when it was just beginning to form bloom buds. How perfectly timed these blooms are, as they open just when the bumblebees that feast on their nectar and pollen appear. Even though the wind is howling, and snow piles up on top of the blossoms, those bees can smell the oddly spring-like perfume of the golden blooms. Remember, though, that according to ancient calendars, spring begins on February 2, the time of first plowing, checking bounds, Imbolc, Groundhog Day, and this numerically magical year, on 2/2/22.

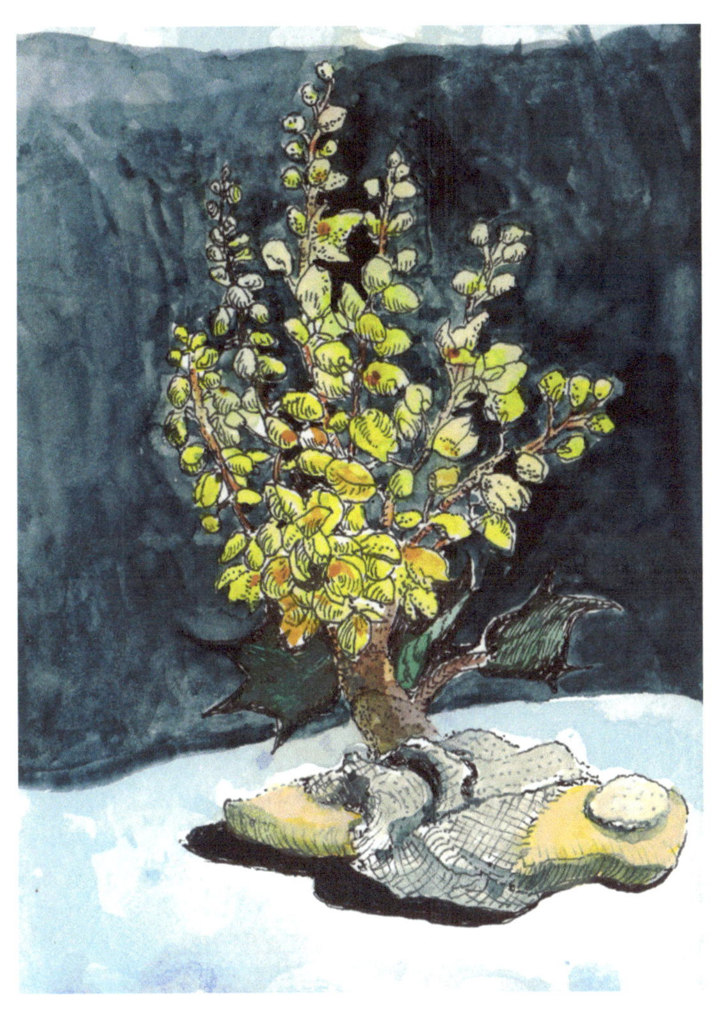

LIVERWORTS, MOSSES, AND LICHENS

In the snowy woods the bright clean greens of the liverworts and mosses and lichens are hard at work giving shelter and food to a few insects, giving oxygen to the air, turning the sunlight into food for the occasional deer. Everything seems to be at a standstill today, breath held at an inhale, waiting for this mid-week to exhale the first soft breath of the start of spring on Wednesday. Imbolc!

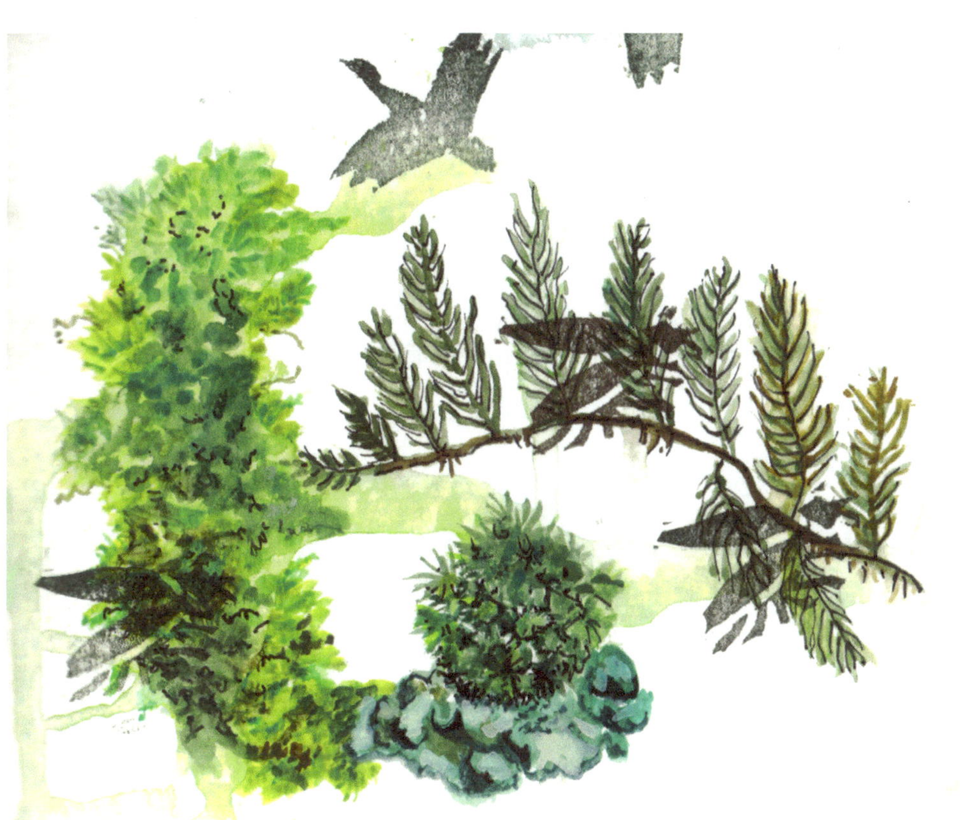

EVERNIA

This little bushy lichen (evernia prunastri, fam. *Parmaliacea*) thrives in clean air up in the sun in the tops of trees (my sample blew down to our street in a recent storm), where it grows very slowly and provides nesting materials for birds and shelter for bugs. It also filters pollutants out of the air and changes carbon dioxide to oxygen. Its presence signifies relatively clean air. During the Industrial Revolution it almost died out in the UK from coal smoke in the air. Today it suffers in farming country from ammonia in the air from manure in fertilizers and from animal slurry.

On a happier note, evernia can survive temperatures below zero, which is why we find it peeking out in snow-covered forests. The sample I drew produced three spore-bearing stalks while I drew it. As a lichen, it has been around for eons. Lichens were one of the earliest forms of life to adapt to living on land. Along with its critical role in nature, this lichen has been appropriated by humans to be of use in the modern perfume industry. It has also been used since ancient times as a lilac-colored dye for fiber.

https://en.wikipedia.org/wiki/Evernia_prunastri

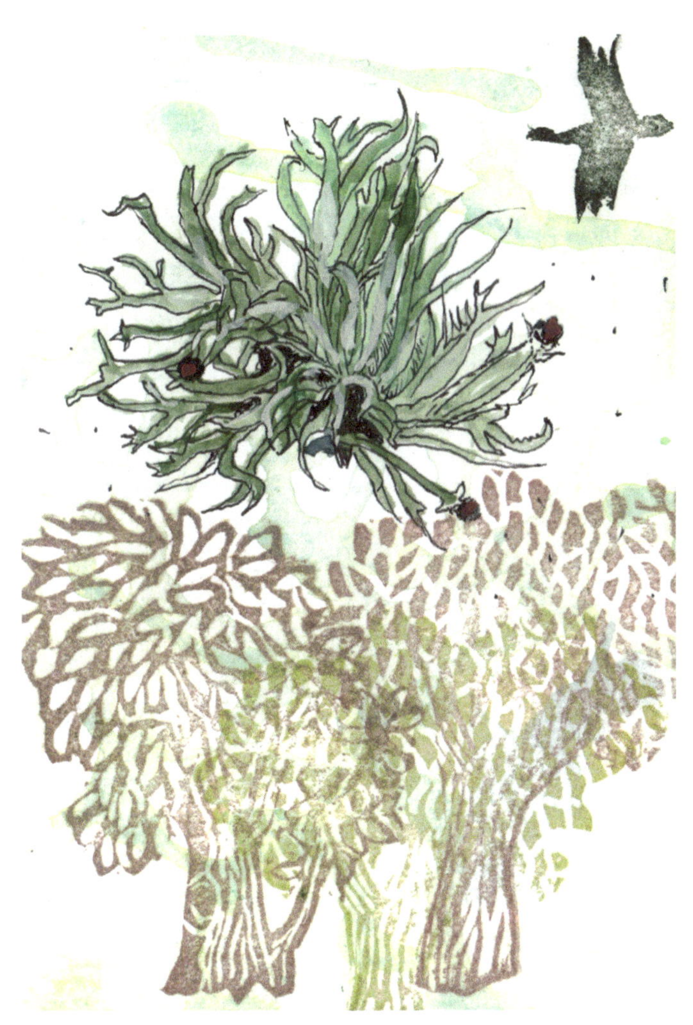

POLYPORES - TUBES

This morning a friend and I walked through some barely managed forest on the edge of a small lake. There were numerous shelf-fungus-covered logs and tree stumps along with moss and lichen-covered live trees. Although I've already written about and drawn shelf fungi (polypores), today I got to see the upward curling undersides of the seashell-like fruiting bodies. These were covered with velvety columns of spore tubes. If you enlarge these drawings, you should be able to see the holes in the tiny tubes. It was above freezing with high humidity this morning, and the polypores seemed to be coming out of their dry state.

The polypores are major actors in the great work of soil building and carbon release. They are most prevalent in old wild, untamed parts of the forest where insects are present to help by boring holes in the decaying wood and feasting on the mycellium-softened interior of the branches.

As you can see here, pale green damp lichens are on the same branch (which had fallen from high up in the tree) photosynthesizing, and they as well as mosses are cleaning the air and replenishing oxygen. I fail to see how management could improve on this small wild place.

http://en.wikipedia.org/wiki/Polypore.

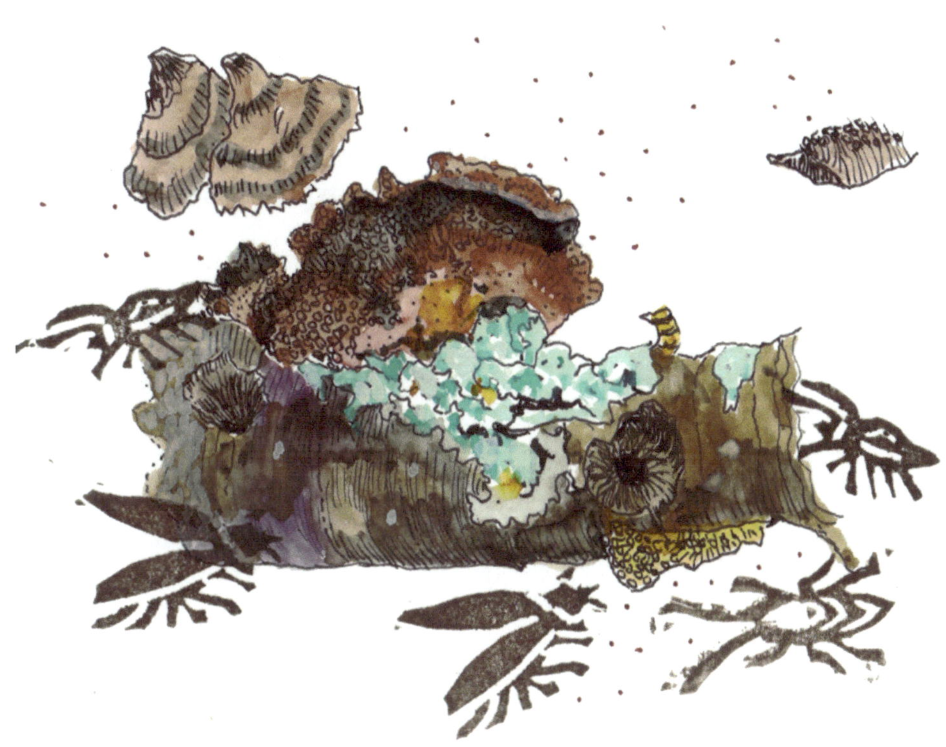

POLYPORES

I came across an entire tree covered with lovely whitish-ochre little shelf or bracket fungi (polypore) in an unmanaged part of the woods the other day. The tops of the shelves, which are the fruiting bodies of this fungus, look like seashells. The growth rings on this group indicates that it is about three to five years old. The undersides house the spore-producing tubes. I expect that the shelves tip upwards when spores and environment are ready for the dispersal.

The fungus itself, the mycellium, is inside the tree, at work recycling the wood and releasing carbon dioxide into the atmosphere for the plants to take in. Polypores (*Polyporaceae* family) are the most important agents of wood decay, an essential process in the forest. They are much more prevalent in unmanaged forests and are the main decomposers of lignin and cellulose.

The fruiting bodies provide shelter and food for numerous insects, and many of them provide food for mammals, including humans. Humans also use them in medicines. If you find some on trees where you live, don't try to remove them from the trees or fallen trunks. They don't set out to kill trees. They grow on trees that are already dead or dying. Their inner fungal parts soften the inner parts of the tree, and insects can then live there; meanwhile, woodpeckers depend on the insects living in the softened wood for their food.

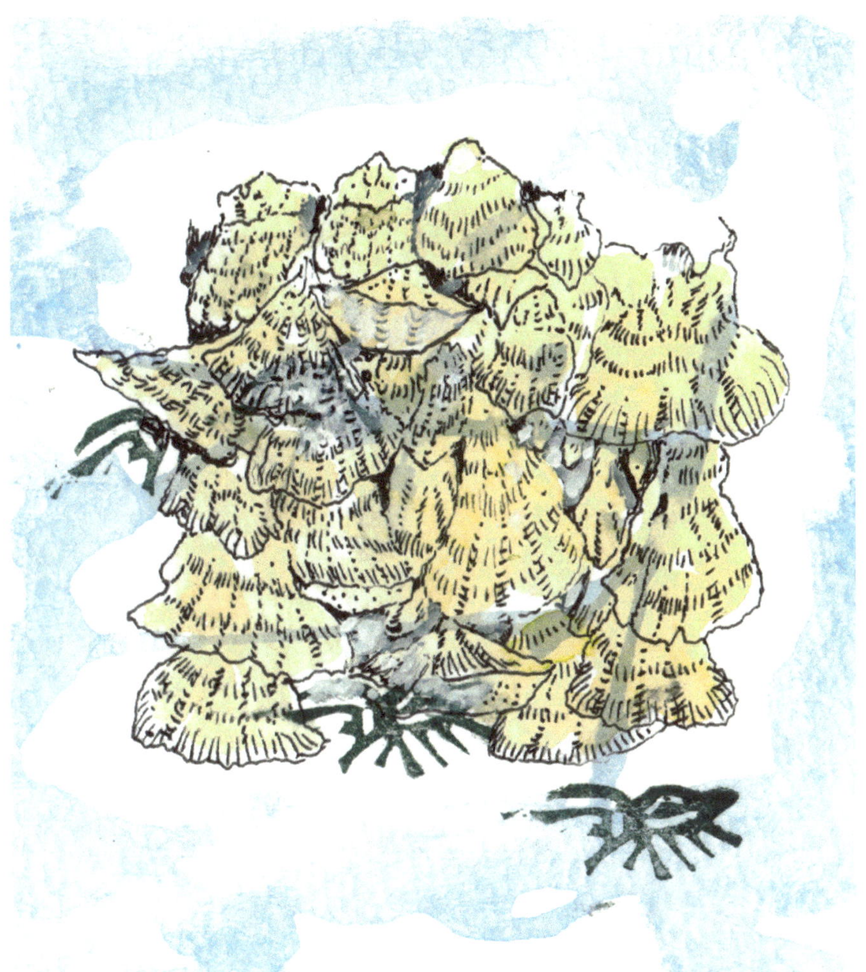

JAPANESE MAGNOLIA -
late winter bloom

The *Magnoleaceae* family is older than bees, butterflies, and moths. Beetles and flies were the only pollinators around at that time. Botanists call these 'dumb pollinators' because of their clumsiness and tendency to chew on the leathery Japanese magnolia leaves in their frenzy to eat the aromatic and nutritious pollen. These days a few bees and stink bugs are also lured in by the aroma, but most arrive too late to pollinate as they chomp their way through the remaining pollen.

The Japanese magnolia trees were lovely this year, covered with the pinkish-lavender blooms that anticipate the leaf buds by several weeks. But suddenly, at the end of a string of warm days, the temperature dropped into the teens on Saturday night, and by 7:00 Sunday morning every blossom was frozen into a tan and brown waxy version of itself, and the beetles had returned underground to their burrows. Each frozen blossom had turned into a protective enclosure around its seed-making structure.

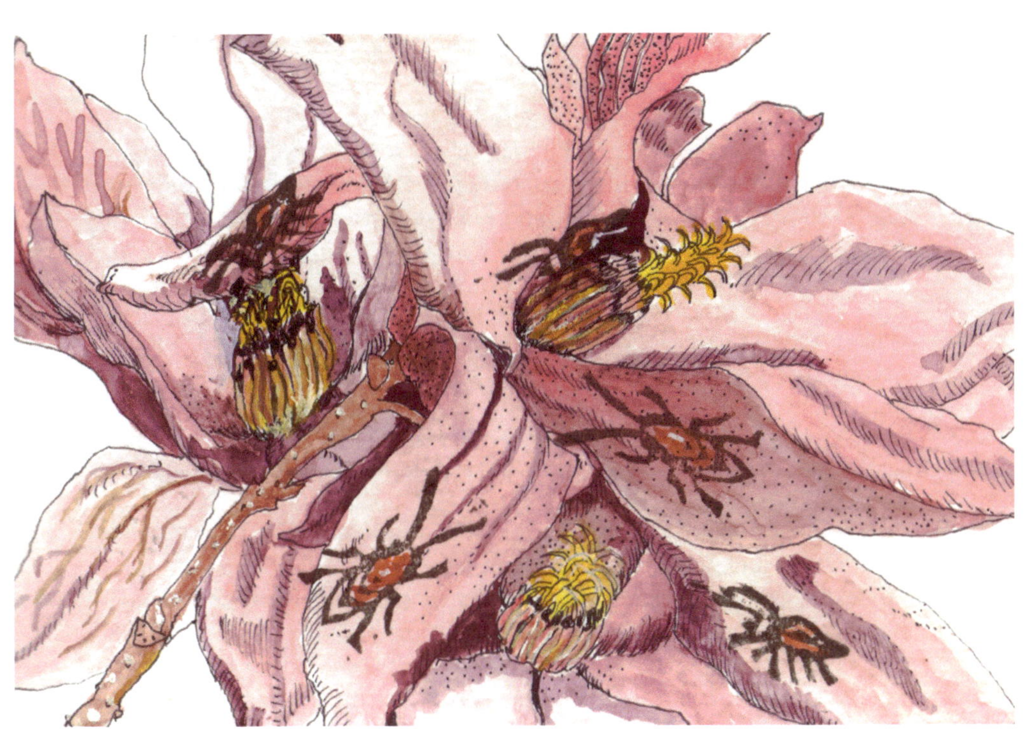

BUSH HONEYSUCKLE

There are a few bush honeysuckles (*Caprifoliaceae* family) along
a fencerow that skirts the east river trail. I love the tender, sweet
smell of the flowers, which I usually smell before I can see them.
They bloom when we are still in danger of frost, so the bloom buds
have pink anthocyanin in places to help them cope with the cold. I
checked out the blooms today, and few were left after last Sunday's
15 degree low; but the tiny green berries seemed to be fine. The
bees had been busy pollinating the blooms for a few weeks.

Bush honeysuckles were brought from temperate Asia to the US to
help humans do erosion control in the late 18th and 19th centuries.
The plants did well here, and birds loved the red berries while deer,
rabbits, and goats ate the leaves, stems, and flowers. Traditional Chinese Medicine has for centuries valued honeysuckle as medicine.

Like so many plants that were brought by humans to this country,
bush honeysuckle is now branded as invasive. Many of the websites
I found suggested poisons to control its spread. Of course the poisons also poison the soil and nearby streams and the adjacent plants
as well as the animals and birds that eat the plants, not to mention
the farm workers who have to do the spraying. In Asian countries
where this plant is native, it is kept in check by the populations of
deer, rabbits, and goats that eat it as well as by careful pruning by
humans.

https://permies.com/t/3865/praise-honey-suckle

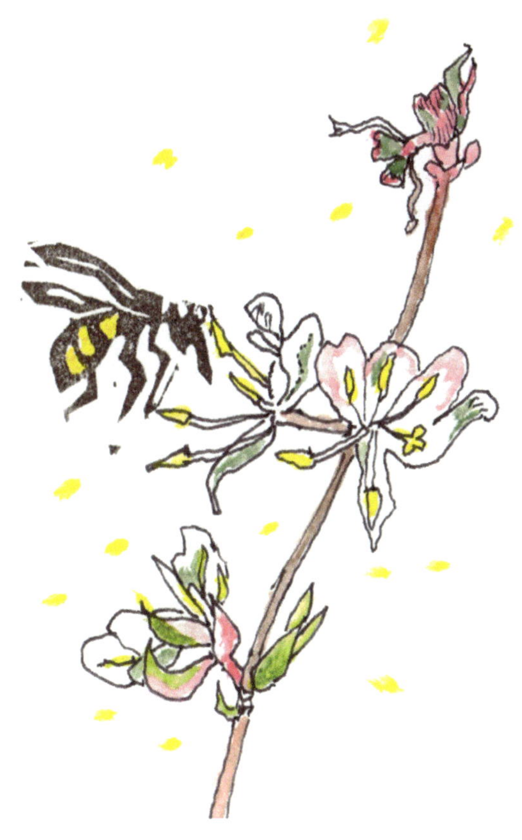

FORSYTHIA AND BEES

Where does the forsythia (*Olivaceae* family) begin and the bee (*Apidae* family) end? What is the thread that ties the bee to the sun to the forsythia to the honey to the human and the bear?

Here we are in what I think of as Yellow Spring. There are so many yellows – forsythia, dandelions, mahonia, daffodils – and they make me wonder why there is so much yellow right now. My acupuncturist tells me that the color yellow is traditionally associated in Chinese Medicine with transitions, nourishment, and earth energies.

From a western scientific perspective, these early blooms are getting their color from carotenoids, pigments that produce the color yellow, which is visible to the bees that pollinate these plants early in the transition from winter to summer. Moreover, carotenoids protect these plants from UV rays, which are strong at this time of year when most leaves have not yet emerged and the flowering plants are exposed to more sunlight. Carotenoids have antibodies that support plant growth and also pollinator health, especially needed during this time of fertilization and seed-making and honey-making. All connected and rhizomatic.

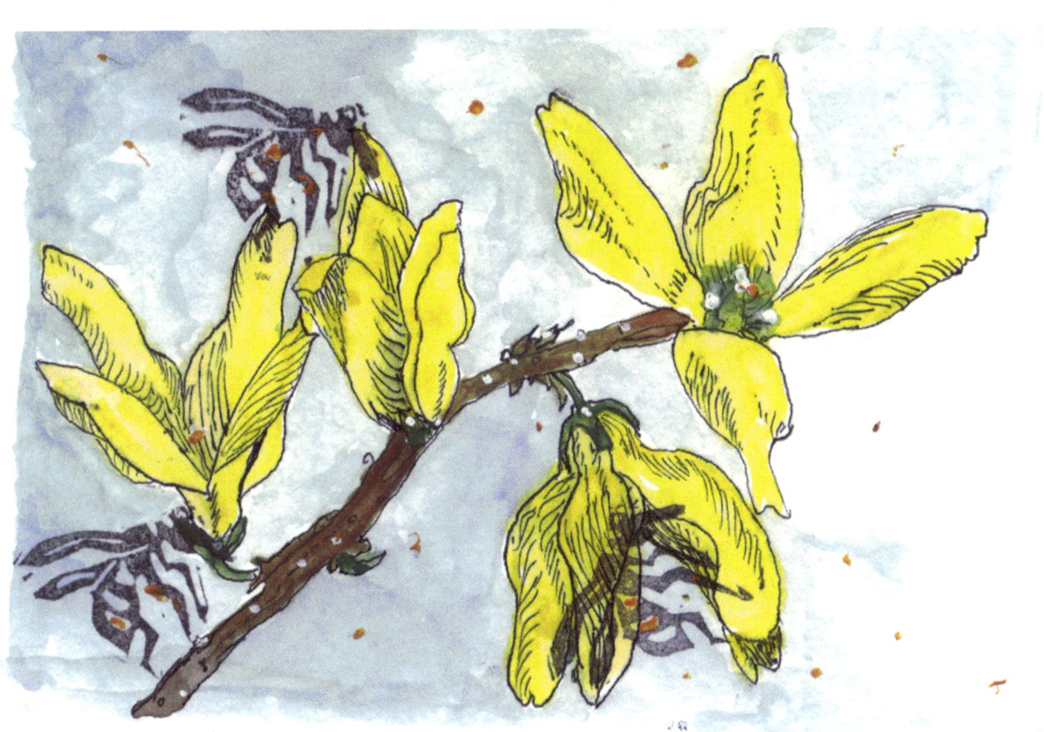

NOTES

NOTES

NOTES